SpringerBriefs in Earth Sciences

For further volumes:
http://www.springer.com/series/8897

Jan Dirk Jansen

A Systems Description of Flow Through Porous Media

 Springer

Jan Dirk Jansen
Department of Geoscience and Engineering
Delft University of Technology
Delft
The Netherlands

ISSN 2191-5369   ISSN 2191-5377   (electronic)
ISBN 978-3-319-00259-0   ISBN 978-3-319-00260-6   (eBook)
DOI 10.1007/978-3-319-00260-6
Springer Cham Heidelberg New York Dordrecht London

Library of Congress Control Number: 2013935769

Printed on acid-free paper

Springer is part of Springer Science+Business Media (www.springer.com)

Preface

This text forms part of material taught during an elective course in advanced reservoir simulation at Delft University of Technology over the past 10 years. The course is aimed at students with an applied mathematics or physics background interested in petroleum engineering applications, and petroleum engineering students seeking to deepen their understanding of the systems and control aspects of reservoir management. Part of the material has also been presented at various short courses for industrial and academic researchers interested in background knowledge needed to perform research in the area of *closed-loop reservoir management*, also known as *smart fields*, related to, e.g., model-based *production optimization*, *data assimilation* (or *history matching*), *model reduction* or *upscaling* techniques. All these topics have connections to system-theoretical concepts such as *observability*, *controllability*, or *identifiability*. In order to facilitate the accessibility of the systems-and-control literature, the course starts with a review of the equations for flow through porous media in a systems-and-control notation. Although the theory is limited to the relatively simple situation of horizontal two-phase (oil–water) flow, it covers several typical aspects of porous media flow. Over the year, various editions of the course notes *"Systems theory for reservoir management"*, which contain much more material than presented in this text, have found their way to interested parties outside the university. The introductory part, i.e., the Systems Description of Flow Through Porous Media, forms the topic of this brief monograph. Chapter or section headings marked with a star superscript indicate material that can be skipped without losing track of the general train of thoughts. I hope the contents of this text will be of value to students and researchers interested in the application of systems and control concepts to oil and gas reservoir simulation and other application areas of subsurface flow simulation such as CO_2 storage, geothermal energy, or groundwater remediation. I like to acknowledge the many students and colleagues, within and outside TU Delft, whose critical comments and constructive remarks have helped to improve the text over the years. If you have any further comments, please let me know at j.d.jansen@tudelft.nl.

Delft, February 2013 — Jan Dirk Jansen

Contents

Symbols

Notes:

- Several symbols occur more than once because they have a different meaning in different parts of the text.
- The dimensions of vectors and matrices have only been indicated when all elements have the same dimensions.

Symbol	Description	Dimensions	SI units
a	Coefficient	–	–
A	Area	L^2	m^2
$\mathbf{A}$	System matrix	–	–
$\overline{\mathbf{A}}$	Jacobian matrix	–	–
b	Coefficient	–	–
B	Oil formation volume factor	–	–
$\mathbf{B}$	Input matrix	–	–
c	Compressibility	$t^2 m^{-1} L^{-1}$	1/Pa
C	Integration constant	$L^{-3} m$	kg/m^3
$\mathbf{C}$	Output matrix	–	–
d	Depth	L	m
D	Diffusion constant	$L^2 t^{-1}$	m^2/s
$\mathbf{D}$	Direct-throughput matrix	–	–
E	Energy	$L^2 m t^{-2}$	J
$\mathbf{E}$	Accumulation matrix	–	–
e	Nonlinear function	–	–
$\mathbf{e}$	Nonlinear vector-valued function	–	–
f	Nonlinear function	–	–
f	Fractional flow	–	–
$\mathbf{f}$	Nonlinear vector-valued function	–	–
$\mathbf{F}$	Fractional-flow matrix	–	–
g	Acceleration of gravity	$L t^{-2}$	m/s^2
g	Velocity gradient	t^{-1}	1/s
$\mathbf{g}$	Nonlinear vector-valued function	–	–
h	Reservoir height	L	m

(continued)

(continued)

h	Nonlinear vector-valued output function	–	–
i	Counter	–	–
i	Unit vector	–	–
j	Counter	–	–
j	Nonlinear vector-valued output function	–	–
I	Identity matrix	–	–
J	Well index, productivity index	L^2m^{-1}	$m^3/$(Pa s)
J	Well index matrix	L^2m^{-1}	$m^3/$(Pa s)
k	Permeability	L^2	m^2
k	Counter	–	–
k	Discrete time	–	–
K	Total number of time steps	–	–
$\vec{\mathbf{K}}$	Permeability tensor	L^2	m^2
L	Length	m	m
L	Spatial differential operator	–	–
L	Location matrix (selection matrix)	–	–
m	Number of elements in input vector **u**	–	–
m	Eigenvector	–	–
M	Matrix of eigenvectors	–	–
n	Number of elements in state vector **x**	–	–
n	Corey exponent	–	–
n	Unit vector normal to boundary	–	–
p	Number of elements in output vector **y**	–	–
p	Pressure	$L^{-1}mt^{-2}$	Pa
p	Pressure vector	$L^{-1}mt^{-2}$	Pa
p	Vector of pressure differences	–	–
P	Power	L^2mt^{-3}	W
q	Flow rate (source term)	L^3t^{-1}	m^3/s
$\tilde{q}$	Flow rate over a grid-block boundary	L^3t^{-1}	m^3/s
q''	Flow rate per unit area (source term)	Lt^{-1}	m/s
q'''	Flow rate per unit volume (source term)	t^{-1}	1/s
q	Vector of flow rates (source terms)	L^3t^{-1}	m^3/s
Q	Scaled flow rate	$L^{-1}mt^{-3}$	Pa/s
r	Radius	L	m
r	Residual	–	–
r	Residual vector	–	–
s	Coordinate along a curve	–	–
s	Saturation vector	–	–
S	Saturation	–	–
S	Matrix to compute $\tilde{v}$ from **p**	$L^2M^{-1}t$	m/(Pa s)
t	Time	t	s
T	Temperature	T	K

(continued)

(continued)

T	Transmissibility	$L^2M^{-1}t$	$m^3/$ (Pa s)
$\mathbf{T}$	Transmissibility matrix	$L^2M^{-1}t$	$m^3/$ (Pa s)
u	Input variable	–	–
$\mathbf{u}$	Input vector	–	–
v	Superficial velocity	Lt^{-1}	m/s
$\tilde{v}$	Interstitial velocity	Lt^{-1}	m/s
$\vec{\mathbf{v}}$	Superficial velocity vector in physical space	Lt^{-1}	m/s
$\mathbf{v}$	Superficial velocity vector	Lt^{-1}	m/s
v	Darcy velocity vector at grid-block boundaries	Lt^{-1}	m/s
$\tilde{v}$	Interstitial velocity vector at grid-block boundaries	Lt^{-1}	m/s
V	Volume	L^3	m^3
$\mathbf{V}$	Accumulation matrix	–	–
x	Spatial coordinate	L	m
x	(state) variable	–	–
$\vec{\mathbf{x}}$	Coordinate vector in physical space	L	m
$\mathbf{x}$	State vector	–	–
y	Spatial coordinate	L	m
y	Output variable	–	–
$\mathbf{y}$	Output vector	–	–
z	Spatial coordinate	L	m
z	Transformed (state) variable	–	–
$\mathbf{z}$	Transformed state vector	–	–
α	Geometric factor	$-, L, L^2$	$-, m, m^2$
α	Valve opening	–	–
$\boldsymbol{\alpha}$	Vector of valve openings	–	–
γ	Vector of geometric factors	L^{-1}	1/m
Γ	Boundary	L^2	m^2
ε	Nonlinear function	–	–
ε	Convergence criterion	–	–
ε	Error	–	–
$\boldsymbol{\varepsilon}$	Vector of model errors	–	–
$\boldsymbol{\eta}$	Vector of measurement errors	–	–
ζ	Diffusion constant	L^2t^{-1}	m^2/s
λ	Mobility	$LM^{-1}t$	$m^2/$ (Pa s)
λ	Eigenvalue	–	–
$\boldsymbol{\lambda}$	Vector of mobilities at grid-block boundaries	$LM^{-1}t$	$m^2/$ (Pa s)
Λ	Diagonal matrix of eigenvalues	–	–
μ	Dynamic viscosity	$L^{-1}mt^{-1}$	Pa s

(continued)

(continued)

π	Dummy variable	–	–
ρ	Density	$L^{-3}m$	kg/m^3
$\Delta\tau$	Grid-block travel time along a stream line	t	s
τ	Time-of-flight along a streamline	t	s
ϕ	Porosity	–	–
Φ	Potential	$L^{-1}mt^{-2}$	Pa
φ	Nonlinear function	–	–
ψ	Source term	–	–
$\boldsymbol{\varphi}$	Vector of averaged grid-block porosities	–	–
Ω	Domain	L^3	m^3
Subscripts			
av	Average		
c	Capillary		
c	Continuous		
con	Connectivity		
d	Discrete		
dis	Dissipation		
D	Dimensionless		
e	Exit		
eq	Equivalent		
gb	Grid-block		
i	Initial		
k	Discrete time		
l	Liquid		
m	Mass		
o	Oil		
op	Oil, pressure		
or	Oil, residual		
os	Oil, saturation		
p	Pore		
p	Pressure		
pot	Potential		
q	Flow rate		
ro	Relative, oil		
rw	Relative, water		
r	Rock		
R	Reservoir		
s	Saturation		
sc	Standard conditions		
$scal$	Scaling		
t	Total		
tf	Flowing tubing-head		
w	Water		
wc	Water, connate		

(continued)

(continued)

wp	Water, pressure
ws	Water, saturation
x	x-Direction
y	y-Direction
λ	Mobility
Superscripts	
0	End-point saturation
0	Linearization point
i	Iteration counter
T	Transpose

Chapter 1
Porous-Media Flow

Abstract This chapter gives a brief review of the basic equations needed to simulate single-phase and two-phase (oil–water) flow through porous media. It discusses the governing partial-differential equations, their physical interpretation (especially the diffusive nature of pressures and the convective behavior of saturations), spatial discretization with finite differences, and the treatment of wells. It contains well-known theory and is primarily meant to form a basis for the next chapter where the equations will be reformulated in terms of systems-and-control notation.

1.1 Introduction

We will restrict the theory to the relatively simple cases of isothermal, slightly compressible single-phase and two-phase (oil–water) flow through porous media, which, however, are sufficient to illustrate many of the typical aspects involved in the numerical simulation of subsurface flow. Moreover, we will only consider spatial discretizations using finite differences, which, however, is no limitation for the development of the theory in subsequent chapters. For more complex flows, involving multiple chemical components with multiple phases and possibly thermal effects and chemical interactions, we refer to standard textbooks on reservoir simulation such as Aziz and Settari (1979), Lake (1989), Mattax and Dalton (1990), or Chen et al. (2006). The latter also treats alternative spatial discretization methods such as finite elements. For further information on mathematical and modeling aspects of reservoir simulation, see e.g. Ewing (1983), Gerritsen and Durlofsky (2005), and Aarnes et al. (2007). For its use in reservoir management, see e.g. Fanchi (2006), Oliver et al. (2008) and Jansen et al. (2008).

J. D. Jansen, *A Systems Description of Flow Through Porous Media*,
SpringerBriefs in Earth Sciences, DOI: 10.1007/978-3-319-00260-6_1,
© The Author(s) 2013

1.2 Notation

Scalars will be indicated with Latin or Greek, lower or upper case letters, and vectors with Latin or Greek lower case letters in **bold-face** or in index notation. Occasionally we will use a ***bold-face-italics*** font to indicate vectors with a special meaning. Matrices will be indicated with Latin or Greek **bold-face** capitals. The superscript T is used to indicate the transpose, and dots above variables to indicate differentiation with respect to time. Unless indicated otherwise, vectors are always considered to be column vectors. For example a vector $\mathbf{x} \in \mathbb{R}^n$ is defined as

$$\mathbf{x} \triangleq \begin{bmatrix} x_1 \\ x_2 \\ \vdots \\ x_n \end{bmatrix} . \tag{1.1}$$

This expression also illustrates the use of the 'embellished' equality sign $\triangleq$ to introduce definitions.

1.3 Single-Phase Flow

1.3.1 Governing Equations

1.3.1.1 General Case

This section gives an overview of the derivation of the governing equations for single-phase flow. For details see e.g. Bear (1972), Peaceman (1977), Aziz and Settari (1979) or Helmig (1997). We consider one-dimensional, horizontal, iso-thermal flow of a compressible single-phase liquid through a compressible porous medium with constant cross-sectional area; see Fig. 1.1.

We can write the mass balance per unit time for a control volume with length dx as:

Fig. 1.1 Control volume

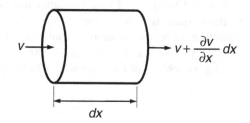

$$\underbrace{A\rho v}_{\text{mass rate in}} - \underbrace{A\left(\rho + \frac{\partial \rho}{\partial x}dx\right)\left(v + \frac{\partial v}{\partial x}dx\right)}_{\text{mass rate out}} - \underbrace{A\frac{\partial \rho \phi}{\partial t}dx}_{\substack{\text{mass accumulated} \\ \text{unit time}}} + \underbrace{\rho q}_{\text{source term}} = 0, \quad (1.2)$$

where A is the cross-sectional area, $\rho(t,x)$ is the fluid density, $v(t,x)$ is the *Darcy velocity* averaged over the cross-section,[1] $\phi(t,x)$ is the porosity, x is the spatial coordinate, t is time and q is a source term. Positive values of q correspond to injection, negative values to production. If we expand Eq. (1.2), drop all terms higher than first order in the differentials, and simplify the results, we obtain

$$\frac{\partial(\rho v)}{\partial x} + \frac{\partial \rho \phi}{\partial t} - \rho q''' = 0, \quad (1.3)$$

where $q'''(t,x)$ is a now a source term expressed as flow rate per unit volume. We can generalize this result to a situation with non-constant cross-section or to two or three dimensions by writing

$$\nabla \cdot (\alpha \rho \vec{v}) + \alpha \frac{\partial(\rho \phi)}{\partial t} - \alpha \rho q''' = 0, \quad (1.4)$$

where $\nabla\cdot$ is the divergence operator, $\alpha(\vec{x})$ is a geometric factor that will be defined below, $\vec{x}$ is the spatial-coordinate vector with components x, y and z, and $\vec{v}(t,\vec{x})$ is the Darcy velocity vector.[2] Depending on whether we consider one-, two- or three-dimensional flow, the factor α, the vectors $\vec{x}$ and $\vec{v}$, and the divergence operator are given by:

$$
\begin{array}{llll}
\text{1–D:} & \alpha = A(x), & \vec{x} = x, & \vec{v} = v, & \nabla\cdot\bullet = \dfrac{\partial\bullet}{\partial x} \\[3mm]
\text{2–D:} & \alpha = h(x,y), & \vec{x} = (x,y), & \vec{v} = (v_x, v_y), & \nabla\cdot\bullet = \dfrac{\partial\bullet}{\partial x} + \dfrac{\partial\bullet}{\partial y} \\[3mm]
\text{3–D:} & \alpha = 1, & \vec{x} = (x,y,z), & \vec{v} = (v_x, v_y, v_z), & \nabla\cdot\bullet = \dfrac{\partial\bullet}{\partial x} + \dfrac{\partial\bullet}{\partial y} + \dfrac{\partial\bullet}{\partial z}
\end{array}
$$

$$(1.5)$$

where h is reservoir height. Conservation of momentum in flow through porous media is usually expressed with Darcy's law, an experimental relationship that

---

[1] The *Darcy velocity* or the *filtration velocity*, is the superficial velocity that would occur if the entire cross section, and not just the pores, would be open to flow. This is as opposed to the *interstitial velocity* $\tilde{v}$, which is defined as $\tilde{v} = v/\phi$, and which is the true fluid velocity in the pore space. The Darcy velocity can also be interpreted as a *volumetric flux*, i.e. the amount of volume flowing through a unit of surface area per unit time.

[2] We use an arrow above a vector or matrix to indicate that it its components are representing quantities in physical space. For example $\vec{v}$ is a velocity vector with one, two or three components, depending on whether we use a one-, two-, or three-dimensional reservoir description. Note that the spatial-coordinate vector $\vec{x}$ is unrelated to the state vector x as used in Chaps. 2 and 3. The use of the same symbol for two different quantities is somewhat unfortunate, but results from conventions in different disciplines.

contains only resistance and gravity terms. Disregarding the inertia forces is justified because of the very low flow velocities that occur in porous-media flow; see Bear (1972). For the one-dimensional, horizontal case with constant cross-sectional area Darcy's law can be expressed as

$$v = -\frac{k}{\mu}\frac{\partial p}{\partial x}, \tag{1.6}$$

where $k(x)$ is the rock *permeability*,[3] and μ is the fluid viscosity. For the more general, non-horizontal, one-, two- or three-dimensional case we can write the same equation in vector form as

$$\vec{v} = -\frac{1}{\mu}\vec{K}(\nabla p - \rho g \nabla d), \tag{1.7}$$

where

$$\nabla \bullet \triangleq \frac{\partial \bullet}{\partial x}, \nabla \bullet \triangleq \left[\frac{\partial \bullet}{\partial x} \ \frac{\partial \bullet}{\partial y}\right]^T \text{ or } \nabla \bullet \triangleq \left[\frac{\partial \bullet}{\partial x} \ \frac{\partial \bullet}{\partial y} \ \frac{\partial \bullet}{\partial z}\right]^T, \tag{1.8}$$

is the gradient operator for one, two or three dimensions respectively, and where $\vec{K}(\vec{x})$ is the permeability tensor, g is the acceleration of gravity and $d(\vec{x})$ is depth. Usually the orientation of the coordinate system can be aligned with the geological layering in the reservoir such that $\vec{K}$ is a diagonal matrix:

$$1\text{--D:} \quad \vec{K} = k, \ 2\text{--D:} \quad \vec{K} = \begin{bmatrix} k_x & 0 \\ 0 & k_y \end{bmatrix}, 3\text{--D:} \quad \vec{K} = \begin{bmatrix} k_x & 0 & 0 \\ 0 & k_y & 0 \\ 0 & 0 & k_z \end{bmatrix}. \tag{1.9}$$

Combining Eqs. (1.4) and (1.7) results in

$$-\nabla \cdot \left[\frac{\alpha\rho}{\mu}\vec{K}(\nabla p - \rho g \nabla d)\right] + \alpha\frac{\partial(\rho\phi)}{\partial t} - \alpha\rho q''' = 0. \tag{1.10}$$

The variables ρ, ϕ, μ and $\vec{K}$ in Eq. (1.10) may all be functions of the pressure p. However, the dependency of μ and $\vec{K}$ on p usually very small and to simplify our formulation we will therefore assume from now on that these parameters are pressure-independent. The relationship between ρ and p follows from the equation of state for a liquid, which can be written in differential form as

---

[3] Permeability has a dimension of length squared and is therefore expressed in SI units in m². In reservoir engineering use is often made of Darcy units, which are defined as: $1 \text{ D} = 9.869233 \times 10^{-13} \approx 10^{-12}$ m².

$$c_l(p) \triangleq \frac{1}{\rho} \frac{\partial \rho}{\partial p}\bigg|_{T_0}, \tag{1.11}$$

where $c_l(p)$ is the isothermal liquid compressibility and T_0 is a constant reference temperature, for which the reservoir temperature T_R is a logical choice. Similarly, the relationship between ϕ and p is given by

$$c_r(p) \triangleq \frac{1}{\phi} \frac{\partial \phi}{\partial p}\bigg|_{T_0}, \tag{1.12}$$

where $c_r(p)$ is the rock compressibility. Equations (1.11) and (1.12) are nonlinear ordinary-differential equations for the dependent variables ρ and ϕ respectively as a function of the independent variable p. They are of first-order and therefore require a boundary condition each, which can be specified as:

$$\rho|_{p=p_0} = \rho_0, \tag{1.13}$$

and

$$\phi|_{p=p_0} = \phi_0. \tag{1.14}$$

With the aid of Eqs (1.11) and (1.12) we can rewrite the accumulation term $\partial(\rho\phi)/\partial t$ in Eq. (1.10) as

$$\frac{\partial(\rho\phi)}{\partial t} = \rho \frac{\partial \phi}{\partial t} + \phi \frac{\partial \rho}{\partial t} = \left(\rho \frac{\partial \phi}{\partial p} + \phi \frac{\partial \rho}{\partial p}\right) \frac{\partial p}{\partial t} = \rho\phi(c_l + c_r) \frac{\partial p}{\partial t} = \rho\phi c_t \frac{\partial p}{\partial t}, \tag{1.15}$$

where $c_t = (c_l + c_r)$ is known as the total compressibility. Combining Eqs. (1.10) and (1.15) results in

$$-\nabla \cdot \underbrace{\left[\frac{\alpha\rho}{\mu} \vec{\mathbf{K}}(\nabla p - \rho g \nabla d)\right]}_{\text{flux term}} + \underbrace{\alpha\rho\phi c_t \frac{\partial p}{\partial t}}_{\text{accumulation term}} - \underbrace{\alpha\rho q'''}_{\text{source term}} = 0. \tag{1.16}$$

Equation (1.16) is a nonlinear partial-differential equation (PDE) for the dependent variable p as a function of the independent variables $\vec{\mathbf{x}}$ and t. It is of first order in t and of second order in $\mathbf{x}$, and therefore requires an initial condition and two boundary conditions for each coordinate direction. The initial condition for p can be written as

$$p(t, \vec{\mathbf{x}})|_{t=t_0} = \breve{p}(\vec{\mathbf{x}}). \tag{1.17}$$

Boundary conditions are usually specified in terms of p (Dirichlet conditions) or $\partial p/\partial \vec{\mathbf{n}}$ (Neumann conditions where $\vec{\mathbf{n}}$ is the outward pointing unit normal vector on the boundary.[4] With the aid of Darcy's law (Eq. (1.7)) the Neumann conditions can be expressed in terms of the velocity, i.e. the flow rate per unit area at the boundary. Therefore we can express the two types of boundary conditions as:

$$p(t, \vec{x})|_{\Gamma_p} = \breve{p}(t) \tag{1.18}$$

and

$$q''(t, \vec{x})|_{\Gamma_q} = \breve{q}''(t), \tag{1.19}$$

where Γ is the boundary of the domain Ω to which Eq. (1.16) applies, and q'' is the outward flow rate per unit area normal to the boundary.

1.3.1.2 Linearized Case

In case of a weakly-compressible liquid and relatively small pressure differences we may assume that ρ and ϕ can be linearized at a reference pressure p_0, while c_l and c_r remain constant. For example to linearize the density, we integrate Eq. (1.11) for constant c_l to obtain

$$\rho = C \exp(c_l p), \tag{1.20}$$

then determine the integration constant C from condition (1.13), use a Taylor expansion to approximate the exponential function, and maintain terms up to first order leading to

$$\rho = \rho_0 \exp[c_l(p - p_0)] \approx \rho_0[1 + c_l(p - p_0)]. \tag{1.21}$$

Similarly, we find for the porosity

$$\phi \approx \phi_0[1 + c_r(p - p_0)]. \tag{1.22}$$

Next, we also assume that the permeability is isotropic, i.e. that the tensor $\vec{K}$ can be replaced by a scalar k, and that α, k and μ are constant over the spatial domain. Substitution of expressions (1.21) and (1.22) in Eq. (1.10), disregarding small terms containing the products $c_l c_r$, $(\nabla p)^2 c_l$ and $\nabla p(p - p_0)c_l$, and dividing out $\alpha \rho_0$ results in the linear equation

$$-\frac{k}{\mu} \nabla^2(p - \rho_0 g d) + \phi_0 c_t \frac{\partial p}{\partial t} - q''' = 0. \tag{1.23}$$

---

[4] More complicated boundary conditions are possible, e.g. by specifying a relationship between p and $\partial p/\partial \mathbf{n}$, a so-called mixed boundary condition. Furthermore, different boundary conditions may be specified along different parts of the boundary.

If we furthermore define the potential $\Phi = p - \rho_0 g d$, Eq. (1.23) can be expressed as a linear diffusion equation

$$\frac{\partial \Phi}{\partial t} = \zeta \nabla^2 \Phi + Q \,, \tag{1.24}$$

where $\zeta = k/(\mu \phi_0 c_t)$ is the diffusion constant, and $Q = q'''/(\phi_0 c_t)$ is a scaled source term.

1.3.2 Finite-Difference Discretization

1.3.2.1 Formulation

This sub-section presents a straightforward approach to the semi-discretization of Eq. (1.16) with the aid of finite differences for 2-D flow. The next sub-section will present an alternative approach that should be used if it is required that the numerical scheme exactly satisfies the mass conservation equations. For further information on finite-difference discretizations, see Peaceman (1977), Aziz and Settari (1979) and Mattax and Dalton (1990). For alternative discretization methods such as finite-volume or finite-element methods, see e.g. Patankar (1980), Helmig (1997), or Chen et al. (2006). First we rewrite Eq. (1.16) in scalar 2-D form, assuming isotropic permeability, small rock and fluid compressibilities, uniform reservoir thickness and absence of gravity forces:

$$-\frac{h}{\mu} \frac{\partial}{\partial x} \left(k \frac{\partial p}{\partial x} \right) - \frac{h}{\mu} \frac{\partial}{\partial y} \left(k \frac{\partial p}{\partial y} \right) + h \phi_0 c_t \frac{\partial p}{\partial t} - h q''' = 0 \,. \tag{1.25}$$

Just like Eq. (1.23), Eq. (1.25) is linear in p. However, unlike Eq. (1.23), Eq. (1.25) does not contain gravity effects, while it still does have the option of a spatial variability of k. Moreover, in Eq. (1.25), we have not divided out the geometric factor h, to stay in line with usual textbook derivation of the discretized equations. Note that because of dividing out the density ρ_0 Eq. (1.25) is now expressed in $m^3 s^{-1}$. We apply a block-centered central-difference scheme with uniform grid to approximate the spatial differentials. The first term in Eq. (1.25) can then be rewritten as

$$\frac{h}{\mu} \frac{\partial}{\partial x} \left(k \frac{\partial p}{\partial x} \right) \approx \frac{h}{\mu} \frac{\Delta}{\Delta x} \left(k \frac{\Delta p}{\Delta x} \right) = \frac{h}{\mu} \frac{k_{i+\frac{1}{2},j} \left(p_{i+1,j} - p_{i,j} \right) - k_{i-\frac{1}{2},j} \left(p_{i,j} - p_{i-1,j} \right)}{(\Delta x)^2} \,, \tag{1.26}$$

where i and j are grid-block counters in x and y direction, and where the subscripts $i + \frac{1}{2}, j$ and $i - \frac{1}{2}, j$ indicate averaged values at the boundaries between grid blocks

Fig. 1.2 One-dimensional
example of harmonic
averaging of grid-block
permeabilities

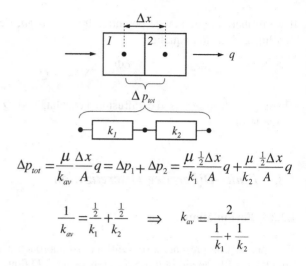

$$\Delta p_{tot} = \frac{\mu}{k_{av}} \frac{\Delta x}{A} q = \Delta p_1 + \Delta p_2 = \frac{\mu}{k_1} \frac{\frac{1}{2}\Delta x}{A} q + \frac{\mu}{k_2} \frac{\frac{1}{2}\Delta x}{A} q$$

$$\frac{1}{k_{av}} = \frac{\frac{1}{2}}{k_1} + \frac{\frac{1}{2}}{k_2} \quad \Rightarrow \quad k_{av} = \frac{2}{\frac{1}{k_1} + \frac{1}{k_2}}$$

(i, j) and $(i + 1, j)$, and grid blocks $(i − 1, j)$ and (i, j) respectively.[5] In analogy to
electrical resistances in series we can work out the series resistance against flow
between two grid-block centers; see Fig. 1.2 for a 1-D example. Similarly, considering flow in the x-direction for the 2-D case we can write

$$\frac{\mu}{h\Delta y} \frac{\Delta x}{k_{i-\frac{1}{2}j}} = \frac{\mu}{h\Delta y} \frac{\frac{1}{2}\Delta x}{k_{i-1,j}} + \frac{\mu}{h\Delta y} \frac{\frac{1}{2}\Delta x}{k_{i,j}} , \tag{1.27}$$

from which we obtain the *harmonic average* for the permeability:

$$k_{i-\frac{1}{2}j} \triangleq \frac{2}{\frac{1}{k_{i-1,j}} + \frac{1}{k_{i,j}}} . \tag{1.28}$$

A similar expression can be obtained for $k_{i+\frac{1}{2}j}$. After rewriting the second term in
Eq. (1.25) in the same fashion, and reorganizing terms we can write

$$Vc(\phi_0)_{i,j} \left(\frac{\partial p}{\partial t}\right)_{i,j} - T_{i-\frac{1}{2}j}p_{i-1,j} - T_{i,j-\frac{1}{2}}p_{i,j-1} + \\ \left(T_{i-\frac{1}{2}j} + T_{i,j-\frac{1}{2}} + T_{i,j+\frac{1}{2}} + T_{i+\frac{1}{2}j}\right)p_{i,j} - T_{i,j+\frac{1}{2}}p_{i,j+1} - T_{i+\frac{1}{2}j}p_{i+1,j} = Vq'''_{i,j}, \tag{1.29}$$

where $V = h\, \Delta x\, \Delta y$ is the grid-block volume (taken identical for all grid blocks)
and where

---

[5] This two-dimensional grid-block numbering is introduced to obtain a systematic description of
the transmissibilities in a two-dimensional reservoir model. In a numerical implementation,
however, one normally uses a one-dimensional grid-block numbering as displayed in Fig. 1.3,
and a *connectivity table* to list the pairs of adjacent grid blocks. See also Table 1.2 which
illustrates the two different numbering systems as applied to Example 1.

$$T_{i-\frac{1}{2},j} \triangleq \frac{\Delta y}{\Delta x} \frac{h}{\mu} k_{i-\frac{1}{2},j} \qquad (1.30)$$

is the transmissibility between grid blocks $(i-1, j)$ and (i, j), with similar expressions for the other transmissibilities. Equation (1.29) can be rewritten in vector form as

$$\begin{bmatrix} 0 & \cdots & 0 & Vc_i(\phi_0)_{i,j} & 0 & \cdots & 0 \end{bmatrix} \begin{bmatrix} \dot{p}_{i,j-1} \\ \vdots \\ \dot{p}_{i-1,j} \\ \dot{p}_{i,j} \\ \dot{p}_{i+1,j} \\ \vdots \\ \dot{p}_{i,j+1} \end{bmatrix}$$

$$+ \begin{bmatrix} -T_{i,j-\frac{1}{2}} & \cdots & -T_{i-\frac{1}{2},j} & \left(T_{i,j-\frac{1}{2}} + T_{i-\frac{1}{2},j} + T_{i+\frac{1}{2},j} + T_{i,j+\frac{1}{2}}\right) & -T_{i+\frac{1}{2},j} & \cdots & -T_{i,j+\frac{1}{2}} \end{bmatrix} \begin{bmatrix} p_{i,j-1} \\ \vdots \\ p_{i-1,j} \\ p_{i,j} \\ p_{i+1,j} \\ \vdots \\ p_{i,j+1} \end{bmatrix} = q_{i,j},$$

$$(1.31)$$

where we have used dots above variables to indicate differentiation with respect to time, and where we have changed from flow rates per unit volume q''' to flow rates q expressed in m³/s. The row vectors in the first and second term of Eq. (1.31) form building blocks for matrices that represent the flow behavior of a collection of grid blocks. The second term of the equation illustrates that, for the chosen 2-D discretization, the change of a grid-block pressure at a certain moment in time is a function of its own value and of the pressure values in the four neighboring grid blocks. The vector with transmissibility matrix elements has therefore typically five non-zero elements. Only the rows that correspond to grid blocks at the edges of the domain require a special treatment to incorporate the boundary conditions. For no-flow boundary conditions this simply means that they only have four non-zero elements, because the fifth one, which represents the boundary transmissibility, is equal to zero. For rows corresponding to grid blocks at a corner of the domain the number of non-zero elements reduces to three. For a system with n grid blocks we can specify n equations of the form (1.31), which, when combined, can be written as

$$\mathbf{V}\dot{\mathbf{p}} + \mathbf{T}\mathbf{p} = \mathbf{q}. \qquad (1.32)$$

The $n \times n$ matrices $\mathbf{T}$ and $\mathbf{V}$ are generally known as the *transmissibility matrix* and the *accumulation matrix* respectively.

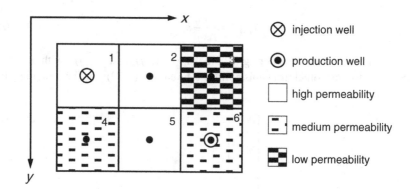

Fig. 1.3 Top view of a six-block finite-difference model of a reservoir with two wells

1.3.3 Example 1: Single-Phase Flow in a Simple Reservoir

We illustrate the structure of the matrices with a (very) simple example. It consists of a finite-difference model of a two-dimensional horizontal reservoir with two vertical wells: an injector in block 1 and a producer in block 6. Figure 1.3 displays the block-centered finite-difference model with six grid blocks. The reservoir and fluid properties have been listed in Table 1.1.

The transmissibility matrix **T** for the six grid blocks of Example 1 can be composed as follows:

Table 1.1 Reservoir and fluid properties for Example 1

Symbol	Variable	Value	SI units	Value	Field units
h	Grid-block height	20	m	66	Ft
$\Delta x, \Delta y$	Grid-block length/width	500	m	1640	Ft
μ	Oil dynamic viscosity	5.0×10^{-4}	Pa s	0.5	cP
k_{low}	Permeability, low	1.0×10^{-14}	m^2	10	mD
k_{med}	Permeability, medium	1.0×10^{-13}	m^2	101	mD
k_{high}	Permeability, high	1.0×10^{-12}	m^2	1013	mD
ϕ	Porosity	0.3	–	0.3	–
c_t	Total compressibility[a]	2.0×10^{-8}	Pa^{-1}	1.4×10^{-4}	psi^{-1}
$\bar{p}_R$	Initial reservoir pressure	30×10^6	Pa	4351.1	psi
r_{well}	Well-bore radius	0.114	m	4.50	In

[a] The value of the compressibility has been chosen about a factor 10 higher than would be expected for an oil-reservoir above bubble-point pressure. This results in additional energy storage in the reservoir, an effect that in reality would occur in the presence of an aquifer and/or a gas cap

Table 1.2 Transmissibilities for Example 1[a]

Connectivity	Grid-block pair	i–j transmissibility numbering	Transmissibility, $m^3/(Pa\ s)$ (10^{-8})
1	1–2	$(1\frac{1}{2}, 1)$	4.000
2	1–4	$(1, 1\frac{1}{2})$	0.727
3	2–3	$(2\frac{1}{2}, 1)$	0.079
4	2–5	$(2, 1\frac{1}{2})$	4.000
5	3–6	$(3, 1\frac{1}{2})$	0.073
6	4–5	$(1\frac{1}{2}, 2)$	0.727
7	5–6	$(2\frac{1}{2}, 2)$	0.727

[a] The two-dimensional $(i$–$j)$ transmissibility numbering in the third column is shown as illustration only. In a numerical implementation one typically uses the connectivities displayed in the first column

$$
\begin{bmatrix}
\left(T_{1,1\frac{1}{2}} + T_{1\frac{1}{2},1}\right) & -T_{1\frac{1}{2},1} & 0 & -T_{1,1\frac{1}{2}} & 0 & 0 \\
-T_{1\frac{1}{2},1} & \left(T_{1\frac{1}{2},1} + T_{2,1\frac{1}{2}} + T_{2\frac{1}{2},1}\right) & -T_{2\frac{1}{2},1} & 0 & -T_{2,1\frac{1}{2}} & 0 \\
0 & -T_{2\frac{1}{2},1} & \left(T_{2\frac{1}{2},1} + T_{3,1\frac{1}{2}}\right) & 0 & 0 & -T_{3,1\frac{1}{2}} \\
-T_{1,1\frac{1}{2}} & 0 & 0 & \left(T_{1,1\frac{1}{2}} + T_{1\frac{1}{2},2}\right) & -T_{1\frac{1}{2},2} & 0 \\
0 & -T_{2,1\frac{1}{2}} & 0 & -T_{1\frac{1}{2},2} & \left(T_{1\frac{1}{2},2} + T_{2,1\frac{1}{2}} + T_{2\frac{1}{2},2}\right) & -T_{2\frac{1}{2},2} \\
0 & 0 & -T_{3,1\frac{1}{2}} & 0 & -T_{2\frac{1}{2},2} & \left(T_{2\frac{1}{2},2} + T_{3,1\frac{1}{2}}\right)
\end{bmatrix}
$$

$$(1.33)$$

Using the data of Table 1.1 we can work out the numerical values of the transmissibilities. The results have been displayed in Table 1.2. The accumulation matrix for the six grid blocks of Example 1 becomes simply

$$
\begin{bmatrix}
Vc_t(\phi_0)_1 & 0 & 0 & 0 & 0 & 0 \\
0 & Vc_t(\phi_0)_2 & 0 & 0 & 0 & 0 \\
0 & 0 & Vc_t(\phi_0)_3 & 0 & 0 & 0 \\
0 & 0 & 0 & Vc_t(\phi_0)_4 & 0 & 0 \\
0 & 0 & 0 & 0 & Vc_t(\phi_0)_5 & 0 \\
0 & 0 & 0 & 0 & 0 & Vc_t(\phi_0)_6
\end{bmatrix} . \quad (1.34)
$$

Equation (1.32) can now be worked out as

$$
10^{-1}
\begin{bmatrix}
0.300 & 0 & 0 & 0 & 0 & 0 \\
0 & 0.300 & 0 & 0 & 0 & 0 \\
0 & 0 & 0.300 & 0 & 0 & 0 \\
0 & 0 & 0 & 0.300 & 0 & 0 \\
0 & 0 & 0 & 0 & 0.300 & 0 \\
0 & 0 & 0 & 0 & 0 & 0.300
\end{bmatrix}
\begin{bmatrix}
\dot{p}_1 \\ \dot{p}_2 \\ \dot{p}_3 \\ \dot{p}_4 \\ \dot{p}_5 \\ \dot{p}_6
\end{bmatrix}
$$

$$
+ \; 10^{-8}
\begin{bmatrix}
4.727 & -4.000 & 0 & -0.727 & 0 & 0 \\
-4.000 & 8.079 & -0.079 & 0 & -4.000 & 0 \\
0 & -0.079 & 0.151 & 0 & 0 & -0.073 \\
-0.727 & 0 & 0 & 1.454 & -0.727 & 0 \\
0 & -4.000 & 0 & -0.727 & 5.454 & -0.727 \\
0 & 0 & -0.073 & 0 & -0.727 & 0.800
\end{bmatrix}
\begin{bmatrix}
p_1 \\ p_2 \\ p_3 \\ p_4 \\ p_5 \\ p_6
\end{bmatrix}
=
\begin{bmatrix}
q_1 \\ 0 \\ 0 \\ 0 \\ 0 \\ q_6
\end{bmatrix} .
$$

$$(1.35)$$

1.3.4 Incompressible Flow

In the special case that the fluid and rock compressibilities are so small that they
may be neglected, it follows from Eq. (1.31) that $\mathbf{V} = \mathbf{0}$. In that case we can
rewrite Eq. (1.32) as

$$\mathbf{Tp} = \mathbf{q} . \tag{1.36}$$

At first sight it appears if Eq. (1.36) can simply be solved for the constant
pressure $\mathbf{p}$. However, as discussed in Sect. 3.1.4, the transmissibility matrix $\mathbf{T}$ is
singular which implies that we cannot directly solve the equation. The singularity
can be removed through prescribing the pressure in at least one of the grid blocks,
or through the use of a well model, in which case we may indeed solve for $\mathbf{p}$ as
discussed in more detail in Sect. 3.2.3.

1.3.5 Mass-Conservative Formulation*

The numerical simulation of a physical process using a discretized form of the
governing PDEs generally results in an approximate solution of those PDEs. In
case of the simulation of reservoir flow this implies that both the mass conservation equation and Darcy's law may not be represented accurately. In reservoir
engineering there is often a desire to adhere to the mass conservation equation as
much as possible, because most simulations are made to predict recoverable
hydrocarbon volumes in some sense. Depending on the discretization used, mass

conservation may be more or less compromised. In discussing this issue we will to a large extent follow the approach of Aziz and Settari (1979), pp. 93–97. The effect of the discretization on the mass-balance error can be understood by considering Eq. (1.31) which can be interpreted as the mass-balance equation for a single grid block. It basically states that the mass accumulation rate of a grid block plus the sum of the mass fluxes to or from the four neighboring grid blocks equals the flow rate of the source term. Note that the matrix coefficients in each row of the transmissibility matrix exactly add up, a property that is also apparent from inspecting the transmissibility matrices in Eqs. (1.33) and (1.35). Because of the symmetry of the transmissibility matrix, the same property holds for each column. Adding the rows of Eq. (1.31) we therefore obtain

$$\sum_{i=1}^{n_x} \sum_{j=1}^{n_y} \left[Vc_t(\phi_0)_{i,j}\dot{p}_{i,j} + q_{i,j} \right], \tag{1.37}$$

where the double sums indicate summation over all grid blocks in the x and y directions, and where the transmissibility terms do no longer appear. Equation (1.37) can be interpreted as the mass-balance equation for the entire system, and it states that the sum of the mass accumulation rates in all grid blocks equals the sum of the source terms. The summation of the source terms does not involve any approximation, and therefore any mass-balance error in the numerical solution results from errors in the accumulation terms $Vc_t \phi_0 \dot{p}$. In our derivation of Eq. (1.25), the starting point for the spatial discretization, we used the assumption of small and constant compressibilities c_l and c_r, and in substituting expressions (1.21) and (1.22) in Eq. (1.10) we therefore disregarded small terms containing the products $c_l c_r$. A straightforward time discretization of the accumulation terms in Eq. (1.31) in the form of

$$Vc_t\phi_0 \frac{\partial p}{\partial t} \approx Vc_t\phi_0 \frac{p_{k+1} - p_k}{\Delta t}, \tag{1.38}$$

where k is the time step indicator, is therefore, in general, not mass-conservative. A mass-conservative time discretization can be obtained by starting from the original form of the accumulation term, $\partial(\rho\phi)/\partial t$, as present in Eq. (1.10). We can now write

$$\frac{\partial(\rho\phi)}{\partial t} \approx \frac{1}{\Delta t} \left[\rho_{k+1}(\phi_{k+1} - \phi_k) + (\rho_{k+1} - \rho_k)\phi_k \right]$$

$$\approx \frac{1}{\Delta t} \left[\rho_{k+1}\phi_0 c_r(p_{k+1} - p_k) + \rho_0 c_l(p_{k+1} - p_k)\phi_k \right] \tag{1.39}$$

$$(\rho_{k+1}\phi_0 c_r + \rho_0\phi_k c_l) \frac{p_{k+1} - p_k}{\Delta t}.$$

In the process of spatially discretizing Eq. (1.25) to arrive at Eq. (1.31) we multiplied with V and divided by ρ_0. Starting from Eq. (1.39), the mass-conservative discretization for the accumulation term in Eq. (1.31) can therefore be written as

$$Vc_t\phi_0 \frac{\partial p}{\partial t} \approx V \left(\frac{\rho_{k+1}}{\rho_0} \phi_0 c_r + \phi_k c_l \right) \frac{p_{k+1} - p_k}{\Delta t} . \tag{1.40}$$

Comparison with Eq. (1.38) shows that the constant coefficient $Vc_t\phi_0$ has been replaced by a state-dependent coefficient, which, moreover, contains an element ρ_{k+1} that should be computed at the new time step $k + 1$. A mass-conservative implementation therefore always requires some form of implicit time integration. For liquid flow, and as long as the pressure changes in the reservoir remain small compared to the total pressure, the effect of mass-balance errors is small, and therefore we do not make use of the strict mass-conserving formulation in our numerical examples. However if compressibility plays a role, e.g. when free gas is present, the use of a mass-conservative formulation is essential.

1.3.6 Well Models

1.3.6.1 Formulation

The flow between two grid blocks is linearly proportional to the product of pressure drop Δp and transmissibility T_{gb}, as was discussed in Sect. 1.3.2 above. However, the pressure close to a well displays very strong, nonlinear, gradients in the radial direction, and to capture this effect accurately a very fine grid around the well is required. Alternatively, one may attempt to model an additional pressure drop based on some analytical or semi-analytical solution for the converging flow around a well. Many authors have treated this topic, see e.g. Aziz and Settari (1979) and the classic paper of Peaceman (1978). For an overview of methods for wells with complex geometries see Ding et al. (2000). Here we will follow Peaceman, who developed an expression for the additional pressure drop due to steady-state cylindrical radial flow towards a well in the center of a grid block. In general, the pressure p as a function of radial distance r from a production well operating at bottom-hole pressure p_{well}, in a homogeneous reservoir with permeability k, and producing fluid with viscosity μ, is given by the logarithmic relationship

$$p = p_{well} - \frac{\mu q}{2\pi kh} \ln \left(\frac{r}{r_{well}} \right) , \tag{1.41}$$

a classic result that follows from solving the steady-state differential equation for radial flow through a porous medium, see e.g. Bear (1972). Note that a negative value of the flow rate q indicates production, and a positive value injection. According to Eq. (1.41) the pressure in an injection well would decrease without limit for increasing r. Similarly, the pressure in a production well would increase without limit. The expression has therefore only physical relevance for a finite domain, bounded by e.g. a circular constant-pressure boundary. Peaceman demonstrated that for the particular case of a repeated five-spot injection-production configuration, the analytical solution for the pressure drop and the numerical solution using a fine grid produce the same value for p at an *equivalent* radial distance from the well in the order of

$$r_{eq} \approx 0.2 \, \Delta x, \tag{1.42}$$

where Δx is the length of the (square) grid blocks. Although this result is only valid for a rather restricted set of assumptions, it has proved to be a very useful basis to model the near-well pressure drop for simple, vertical, wells in regular grids. In a follow-up paper, Peaceman argued that for rectangular grid blocks with length Δx and width Δy, expression (1.42) should be modified to

$$r_{eq} = 0.14 \sqrt{\Delta x^2 + \Delta y^2}, \tag{1.43}$$

see Peaceman (1983). Combining Eqs. (1.41) and (1.43) we find for the additional pressure drop between the grid-block pressure and the well-bottom-hole pressure

$$p_{gb} - p_{well} = -\frac{q}{J_{well}} = -\frac{\mu q}{2\pi kh} \ln\left(\frac{0.14\sqrt{\Delta x^2 + \Delta y^2}}{r_{well}}\right), \tag{1.44}$$

where J_{well} is known as the *well index* or *productivity index*, and where negative flow rates indicate production.

1.3.6.2 Example 1: Well Model

If we consider the wells in our Example 1 and fill in the numerical values of Table 1.1 we find that $J_{well,11} = 3.72 \times 10^{-8}$ m^3/(Pa s) and $J_{well,66} = 3.72 \times 10^{-9}$ m^3/(Pa s).

1.4 Two-Phase Flow

1.4.1 Governing Equations

This section gives a brief overview of the derivation of the governing PDEs for two-phase (oil–water) flow, using the *simultaneous solution method* formulated in p_o and S_w as described in Aziz and Settari (1979), p. 133. We consider isothermal conditions only and we will formulate the equations in terms of in situ volumes. The often applied formulation in terms of surface volumes, using formation volume factors, is not necessary for our purpose. The mass-balance equations can be expressed for each of the phases as

$$\nabla \cdot (\alpha \rho_w \vec{v}_w) + \alpha \frac{\partial (\rho_w \phi S_w)}{\partial t} - \alpha \rho_w q_w''' = 0 \, , \tag{1.45}$$

$$\nabla \cdot (\alpha \rho_o \vec{v}_o) + \alpha \frac{\partial (\rho_o \phi S_o)}{\partial t} - \alpha \rho_o q_o''' = 0 \, , \tag{1.46}$$

where subscripts w an o are used to identify water and oil, and where S_w and S_o are the saturations, defined as the fraction of the pore space occupied by the respective phase. Darcy's law can now be expressed as

$$\vec{v}_w = -\frac{k_{rw}}{\mu_w} \vec{\mathbf{K}} (\nabla p_w - \rho_w g \nabla d) \, , \tag{1.47}$$

$$\vec{v}_o = -\frac{k_{ro}}{\mu_o} \vec{\mathbf{K}} (\nabla p_o - \rho_o g \nabla d) \, , \tag{1.48}$$

where k_{rw} and k_{ro} are the relative permeabilities, which represent the additional resistance to flow of a phase caused by the presence of the other phase. For an explanation of the underlying physical concepts, see e.g. Lake (1989). Combining Eqs. (1.45) to (1.48) we obtain:

$$-\nabla \cdot \left[\frac{\alpha \rho_w k_{rw}}{\mu_w} \vec{\mathbf{K}} (\nabla p_w - \rho_w g \nabla d) \right] + \alpha \frac{\partial (\rho_w S_w \phi)}{\partial t} - \alpha \rho_w q_w''' = 0 \, , \tag{1.49}$$

$$-\nabla \cdot \left[\frac{\alpha \rho_o k_{ro}}{\mu_o} \vec{\mathbf{K}} (\nabla p_o - \rho_o g \nabla d) \right] + \alpha \frac{\partial (\rho_o S_o \phi)}{\partial t} - \alpha \rho_o q_o''' = 0 \, . \tag{1.50}$$

Equations (1.49) and (1.50) together contain four unknowns, p_w, p_o S_w and S_o, two of which can be eliminated with aid of the relationships

$$S_w + S_o = 1 \, , \tag{1.51}$$

$$p_o - p_w = p_c(S_w) \, , \tag{1.52}$$

where $p_c(S_w)$ is the oil–water capillary pressure. Substituting Eqs. (1.51) and (1.52) in Eqs. (1.49) and (1.50), expanding the right-hand sides, applying chain-rule differentiation, and substituting the isothermal equations of state

$$c_o \triangleq \frac{1}{\rho_o} \frac{\partial \rho_o}{\partial p_o} \bigg|_{T_R} \, , \tag{1.53}$$

$$c_w \triangleq \frac{1}{\rho_w} \frac{\partial \rho_w}{\partial p_w} \bigg|_{T_R} \approx \frac{1}{\rho_w} \frac{\partial \rho_w}{\partial p_o} \bigg|_{T_R} \, , \tag{1.54}$$

and the definition of rock compressibility

$$c_r \triangleq \frac{1}{\phi}\frac{\partial \phi}{\partial p_o}, \tag{1.55}$$

allows us to express Eq. (1.49) in terms of p_o and S_w as follows:

$$-\nabla \cdot \left\{ \frac{\alpha \rho_w k_{rw}}{\mu_w} \vec{\mathbf{K}} \left[\left(\nabla p_o - \frac{\partial p_c}{\partial S_w} \nabla S_w \right) - \rho_w g \nabla d \right] \right\} + \\ \alpha \rho_w \phi \left[S_w (c_w + c_r) \frac{\partial p_o}{\partial t} + \frac{\partial S_w}{\partial t} \right] - \alpha \rho_w q_w''' = 0, \tag{1.56}$$

$$-\nabla \cdot \left[\frac{\alpha \rho_o k_{ro}}{\mu_o} \vec{\mathbf{K}} (\nabla p_o - \rho_o g \nabla d) \right] + \\ \alpha \rho_o \phi \left[(1 - S_w)(c_o + c_r) \frac{\partial p_o}{\partial t} - \frac{\partial S_w}{\partial t} \right] - \alpha \rho_o q_o''' = 0. \tag{1.57}$$

The diffusive effect of capillary pressure plays a role during displacement processes on a relatively small length scale (as e.g. in core-flooding experiments). During water flooding on reservoir scale the dispersive effect of geological heterogeneities is usually much larger than the diffusive effect of capillary pressures. The correct way to take this dispersion into account is through the use of a velocity-dependent dispersion tensor; see Russell and Wheeler (Russel and Wheeler 1983). In addition to diffusion and dispersion caused by physical phenomena, artificial diffusion will occur as a result of the numerical solution of the discretized form of the equations. In many cases this numerical diffusion is of the same order of magnitude as or even larger than the physical diffusion and dispersion. At this point we will simply neglect capillary forces and dispersion altogether. Equations (1.56) and (1.57) can then be simplified to:

$$-\nabla \left[\frac{\alpha \rho_w k_{rw}}{\mu_w} \vec{\mathbf{K}} (\nabla p - \rho_w g \nabla d) \right] + \alpha \rho_w \phi \left[S_w (c_w + c_r) \frac{\partial p}{\partial t} + \frac{\partial S_w}{\partial t} \right] - \alpha \rho_w q_w''' = 0, \tag{1.58}$$

$$-\nabla \left[\frac{\alpha \rho_o k_{ro}}{\mu_o} \vec{\mathbf{K}} (\nabla p - \rho_o g \nabla d) \right] + \alpha \rho_o \phi \left[(1 - S_w)(c_o + c_r) \frac{\partial p}{\partial t} - \frac{\partial S_w}{\partial t} \right] - \alpha \rho_o q_o''' \\ = 0, \tag{1.59}$$

where the subscript 'o' has been dropped for the pressure because the absence of capillary pressure implies that $p_o = p_w$.

1.4.2 Nature of the Equations

The nature of two-phase flow equations is discussed by e.g. Peaceman (1977),
Aziz and Settari (1979), Ewing (1983) and Lake (1989). They illustrate that the
pressure behavior is essentially diffusive, i.e. that the corresponding equations are
parabolic and become elliptic in the limit of zero compressibility. The saturation
behavior is mixed diffusive-convective, i.e. the corresponding equations are mixed
parabolic-hyperbolic and become completely hyperbolic in the case of zero capillary pressure. This can be seen by rewriting the equations as follows. Consider
Eqs. (1.56) and (1.57) for 1-D flow through a conduit with constant cross-sectional
area A, for small compressibilities such that we may assume that ρ is constant but
c finite, in the absence of gravity terms and capillary pressure and source terms,[6]
and with isotropic permeability k:

$$-\frac{\partial}{\partial x}\left(\lambda_w \frac{\partial p}{\partial x}\right) + \phi\left[S_w(c_w + c_r)\frac{\partial p}{\partial t} + \frac{\partial S_w}{\partial t}\right]_w = 0 , \qquad (1.60)$$

$$-\frac{\partial}{\partial x}\left(\lambda_o \frac{\partial p}{\partial x}\right) + \phi\left[(1 - S_w)(c_o + c_r)\frac{\partial p}{\partial t} - \frac{\partial S_w}{\partial t}\right] = 0 . \qquad (1.61)$$

Here we introduced the water and oil mobilities

$$\lambda_w \triangleq \frac{kk_{rw}(S_w)}{\mu_w} \qquad (1.62)$$

and

$$\lambda_o \triangleq \frac{kk_{ro}(S_w)}{\mu_o} . \qquad (1.63)$$

Addition of Eqs. (1.60) and (1.61) results in a PDE with only the pressure as
primary variable[7]

$$-\frac{\partial}{\partial x}\lambda_t \frac{\partial p}{\partial x} + \phi c_t \frac{\partial p}{\partial t} = 0 , \qquad (1.64)$$

where the total mobility λ_t, and the total compressibility c_t have been defined as

$$\lambda_t \triangleq \lambda_w + \lambda_o , \qquad (1.65)$$

$$c_t \triangleq S_w c_w + (1 - S_w)c_o + c_r . \qquad (1.66)$$

---

[6] Absence of source terms corresponds to considering the (1-D) flow between an injector and a
producer, in which case the flow is driven through the boundary conditions.

[7] The coefficients are still functions of saturation.

Equation (1.64) is a parabolic equation with non-constant coefficients. In the incompressible case the equation reduces to an elliptic equation:

$$\frac{\partial}{\partial x} \lambda_t \frac{\partial p}{\partial x} = 0.$$ (1.67)

Another equation, with only the water saturation as primary variable, can be obtained as follows. Neglecting gravity and considering 1-D flow, Darcy's law for water and oil, as given in Eqs. (1.47) and (1.48), can be expressed as

$$v_w = -\lambda_w \frac{\partial p}{\partial x},$$ (1.68)

$$v_o = -\lambda_o \frac{\partial p}{\partial x}.$$ (1.69)

Furthermore, we make use of the ratio

$$f_w \triangleq \frac{v_w}{v_w + v_o} = \frac{\lambda_w}{\lambda_w + \lambda_o},$$ (1.70)

known as the water *fractional flow*, where $v_w + v_o$ represents the *total velocity* v_t. With the aid of these expressions, and realizing that f_w is a function[8] of S_w and therefore that $\partial f_w / \partial x = (\partial f_w / \partial S_w)(\partial S_w / \partial x)$, we can rewrite Eqs. (1.60) and (1.61) as

$$v_t \frac{\partial f_w}{\partial S_w} \frac{\partial S_w}{\partial x} + \phi \left[S_w(c_w + c_r) \frac{\partial p}{\partial t} + \frac{\partial S_w}{\partial t} \right] = 0,$$ (1.71)

$$-v_t \frac{\partial f_w}{\partial S_w} \frac{\partial S_w}{\partial x} + \phi \left[(1 - S_w)(c_o + c_r) \frac{\partial p}{\partial t} - \frac{\partial S_w}{\partial t} \right] = 0.$$ (1.72)

Subtraction of Eqs. (1.71) and (1.72) after premultiplication with the appropriate factors allows us to eliminate the $\partial p / \partial t$ term and to obtain the required equation in terms of saturations only. In particular, for the incompressible case we find that:

$$v_t \frac{\partial f_w}{\partial S_w} \frac{\partial S_w}{\partial x} + \phi \frac{\partial S_w}{\partial t} = 0.$$ (1.73)

Equation (1.73) is a first-order nonlinear hyperbolic equation, with a non-constant coefficient $v_t = v_w + v_o$ which depends on the pressure according to Eqs. (1.68) and (1.69). In theory, the coupled Eqs. (1.64) and (1.73) are therefore both nonlinear. However, because the changes in saturations usually occur much slower

---

[8] $f_w(S_w)$ is sometimes referred to as the *flux function*.

than the pressure changes,[9] the nonlinearity in the pressure Eq. (1.64) is weak, and the equation may often be considered as a linear one with slowly-varying coefficients.

1.4.3 Relative Permeabilities

The saturation-dependency of the relative permeabilities is usually determined from laboratory experiments where water is forced through a small *core* (a cylindrical piece of rock) initially saturated with oil. The values to be used in reservoir simulation are typically provided in the form of tables or simple mathematical expressions with parameters that have been fitted using the experimental results. Several of such expressions are known in the literature. Here we use the so-called Corey model given by

$$k_{rw} = k_{rw}^0 S^{n_w}, \tag{1.74}$$

$$k_{ro} = k_{ro}^0 (1 - S)^{n_o}, \tag{1.75}$$

where S is a scaled saturation defined as

$$S \triangleq \frac{S_w - S_{wc}}{1 - S_{or} - S_{wc}}, \, 0 \leq S \leq 1, \tag{1.76}$$

k_{rw}^0 and k_{ro}^0 are the *end-point relative permeabilities*, n_w and n_o are the *Corey exponents*, S_{wc} is the *connate-water saturation* and S_{or} is the *residual-oil saturation*. Note that the water fractional flow can also be expressed as

$$f_w = \frac{\lambda_w}{\lambda_w + \lambda_o} = \frac{\frac{k_{rw}}{\mu_w}}{\frac{k_{rw}}{\mu_w} + \frac{k_{ro}}{\mu_o}} = \frac{k_{rw}}{k_{rw} + k_{ro} \frac{\mu_w}{\mu_o}}. \tag{1.77}$$

Moreover, in the next section we will make use of the derivative df_w/dS_w which can, for the Corey model, be expressed analytically as

$$\frac{df_w}{dS_w} = \frac{\frac{dk_{rw}}{dS_w}}{k_{rw} + k_{ro} \frac{\mu_w}{\mu_o}} - \frac{k_{rw} \left(\frac{dk_{rw}}{dS_w} + \frac{dk_{ro}}{dS_w} \frac{\mu_w}{\mu_o} \right)}{\left(k_{rw} + k_{ro} \frac{\mu_w}{\mu_o} \right)^2}, \tag{1.78}$$

---

[9] In fact, the hyperbolic saturation equation in the form given in Eq. (1.73) is coupled to the elliptic pressure Eq. (1.67) because we assumed incompressible flow in its derivation. In that case the pressure changes are instantaneous.

where

$$\frac{dk_{rw}}{dS_w} = \frac{k_{rw}^0 n_w S^{n_w-1}}{1 - S_{wc} - S_{or}} ,$$

(1.79)

$$\frac{dk_{ro}}{dS_w} = -\frac{k_{ro}^0 n_o (1 - S)^{n_o-1}}{1 - S_{wc} - S_{or}} .$$

(1.80)

1.4.4 Example 2: Two-Phase Flow in a Simple Reservoir

Example 2 consists of the same six-block reservoir as in Example 1, but with two fluids, oil and water, instead of just oil. The additional reservoir and fluid properties have been specified in Table 1.3, and the corresponding relative permeabilities and the water fractional flow have been plotted in Figs. 1.4 and 1.5. The water and oil compressibilities are equal and identical to the oil compressibility from Example 1, such that the total compressibility also remains unchanged. In this particular case the accumulation terms are therefore not a function of the saturations. The water viscosity is twice the oil viscosity. Moreover, the end-point permeability of water is two-thirds of the end-point permeability of oil, such that the end-point water–oil mobility ratio is equal to one-third, i.e. favorable. Figures 1.4 and 1.5 clearly display the strong saturation dependency of the relative permeabilities and the corresponding water fractional flow.

1.4.5 Buckley-Leverett Equation*

Equation (1.73) is often referred to as the *Buckley-Leverett* equation after the authors who first presented and analyzed it in the petroleum literature (Buckley and Leverett 1942). It describes the one-dimensional saturation distribution of two incompressible immiscible fluids, neglecting the effects of gravity and capillary pressure. Without loss of generality consider a core-flooding experiment with boundary and initial conditions given by

Table 1.3 Additional reservoir and fluid properties for Example 2

Symbol	Variable	Value	SI units	Value	Field units
μ_w	Water dynamic viscosity	1.0×10^{-3}	Pa s	1.0	cP
k_{ro}^0	End-point relative permeability, oil			0.9	–
k_{rw}^0	End-point relative permeability, water			0.6	–
n_o	Corey exponent, oil			2.0	–
n_w	Corey exponent, water			2.0	–
S_{or}	Residual-oil saturation			0.2	–
S_{wc}	Connate-water saturation			0.2	–

Fig. 1.4 Relative
permeabilities for Example 2

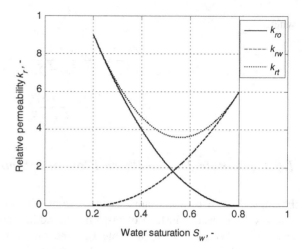

Fig. 1.5 Water fractional
flow for Example 2

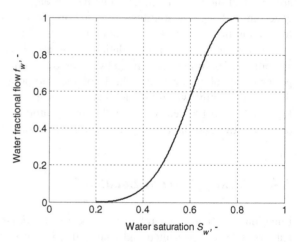

$$S_w(x,0) = S_{wc} , \tag{1.81}$$

$$S_w(0,t) = 1 - S_{or} , \tag{1.82}$$

representing a situation where the core is initially filled with oil except for a small fraction S_{wc} of connate water, whereafter water flooding takes place by injecting water at $x = 0$ such that the oil is replaced by water except for a small fraction S_{ro} of residual oil. As for all hyperbolic equations (which typically describe wave propagation problems) it is possible to find *characteristics*, i.e. relationships between x and t for which the dependent variables do not change. In our case of a single dependent variable S_w this means that the total derivative dS_w/dt should remain equal to zero:

$$\frac{dS_w}{dt} = \frac{\partial S_w}{\partial t} + \frac{\partial S_w}{\partial x}\frac{\partial x}{\partial t} = 0 . \tag{1.83}$$

For a given saturation $S_w = \hat{S}_w$ we can therefore write

$$\left.\frac{dx}{dt}\right|_{S_w=\hat{S}_w} = \frac{\frac{\partial S_w}{\partial t}\big|_{S_w=\hat{S}_w}}{\frac{\partial S_w}{\partial x}\big|_{S_w=\hat{S}_w}} \, , \tag{1.84}$$

and combination of Eqs. (1.73) and (1.84) then gives

$$\left.\frac{dx}{dt}\right|_{S_w=\hat{S}_w} = \frac{v_t}{\phi} \left.\frac{df_w}{dS_w}\right|_{S_w=\hat{S}_w} . \tag{1.85}$$

The position of the point where $S_w = \hat{S}_w$ follows by integrating Eq. (1.85) resulting in

$$\left.x\right|_{S_w=\hat{S}_w} = \frac{v_t\, t}{\phi} \left.\frac{df_w}{dS_w}\right|_{S_w=\hat{S}_w} \, , \tag{1.86}$$

where the integration constant has been set equal to zero which implies that $x = 0$ at $t = 0$. If the core has length L and cross sectional area A, it is convenient to rescale Eq. (1.86) as

$$\left.x_D\right|_{S_w=\hat{S}_w} = t_D \left.\frac{df_w}{dS_w}\right|_{S_w=\hat{S}_w} \, , \tag{1.87}$$

which leads to the dimensionless Buckley-Leverett velocity

$$\left.v_D\right|_{S_w=\hat{S}_w} = \left.\frac{df_w}{dS_w}\right|_{S_w=\hat{S}_w} . \tag{1.88}$$

Here the dimensionless length and time are defined as

$$x_D \triangleq \frac{x}{L} \, , \tag{1.89}$$

$$t_D \triangleq \frac{A v_t t}{A L \phi} = \frac{q_t t}{V_p} \, , \tag{1.90}$$

where V_p is the pore volume of the core. Figure 1.6 displays the derivative df_w/dS_w as a function of S_w, and Fig. 1.7 the corresponding Buckley-Leverett solution (1.87) at dimensionless time $t_D = 0.2$, i.e. after injection of 20 % of the pore volume or $(1 - S_{or} - S_{wc}) \times 20\% = 33.3\%$ of the mobile pore volume.

Figure 1.7 illustrates that the Buckley-Leverett solution predicts the existence of three values of S_w over a large part of the core (for the example of $t_D = 0.2$ this occurs for values of x_D below 0.73), a situation that is clearly physically impossible. The unphysical solutions originate from neglecting the effect of capillary pressure which in reality produces a sharp increase in water saturation at a value of x_D somewhere in the triple value region. This effect can be approximated in the form of a step-wise increase of saturation known as a *shock* in the theory of hyperbolic differential equations. The magnitude and position of the shock follow

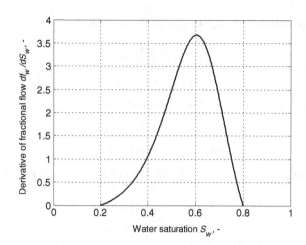

Fig. 1.6 Derivative of the water fractional flow for Example 2

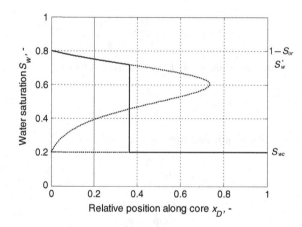

Fig. 1.7 Buckley-Leverett solution (*dotted line*) and shock solution (*solid line*) corresponding to $t_D = 0.2$ or 33.3 % mobile pore volume injected

from requiring that mass is conserved in a control volume around the shock.[10] If the shock of magnitude $\Delta S_w = S_w^* - S_{wc}$ moves a distance Δx_D in a time interval Δt_D it can be derived that the *shock velocity* v_D^* should obey (see e.g. Lake (1989) for details)

---

[10] We may either consider the water mass or the oil mass. Moreover, because we assume the fluids to be incompressible, it is sufficient to consider a volume balance rather than a mass balance.

$$v_D^* = \frac{\Delta x_D}{\Delta t_D} = \frac{f(S_w^*)}{S_w^* - S_{wc}} , \qquad (1.91)$$

where we used a superscripted star to indicate the variables at shock conditions. However, the velocity should also satisfy Eq. (1.88), and we may therefore equate expressions (1.88) and (1.91) which leads to a condition for the shock saturation S_w^* in the form of

$$\left.\frac{df_w}{dS_w}\right|_{S_w=S_w^*} = \frac{f(S_w^*)}{S_w^* - S_{wc}} . \qquad (1.92)$$

Usually Eq. (1.92) cannot be solved explicitly for S_w^*, and requires an iterative numerical solution.[11] The solid line in Fig. 1.7 displays the saturation profile along the core taking into account the shock formation. Combining Eqs. (1.87) and (1.91) we can express the full solution as

$$x_D(S_w, t_D) = \begin{cases} \frac{df_w}{dS_w} t_D, & S_w^* \leq S_w \leq 1 - S_{or} \\ v_D^* t_D, & S_{wc} \leq S_w \leq S_w^* \end{cases} . \qquad (1.93)$$

This kind of analysis, known as *fractional-flow theory* or the *method of characteristics*, has been successfully extended to multiple components and thermal behavior, see e.g. Lake (1989), but is typically restricted to one-dimensional flow. Moreover, in reality there will always be dispersive effects caused by capillary forces, compressibility and reservoir heterogeneities, and true shocks will therefore never be present. However, sharp saturation increases do certainly occur and the underlying characteristic hyperbolic behavior of the saturation equation is an important feature of multiphase flow in porous media.

1.4.6 Linear Approximation*

In the special case of linear relative permeabilities with end values equal to one and residual saturations equal to zero we have

$$k_{rw}(S_w) = S_w , \qquad (1.94)$$

and

$$k_{ro}(S_w) = 1 - S_w . \qquad (1.95)$$

If, in addition, we take

---

[11] As was first shown in Welge (1952), Eq. (1.92) implies that the tangent at f_w in S_w^* and the secant from S_{wc} to S_w^* are identical, leading to a simple graphical solution procedure, known as the *Welge method*, which was popular before the advent of computers.

$$\mu \overset{\triangle}{=} \mu_o = \mu_w \,, \tag{1.96}$$

we have

$$\lambda \overset{\triangle}{=} \frac{k}{\mu} \,, \tag{1.97}$$

and

$$f_w = S_w, \tag{1.98}$$

such that we can rewrite Eqs. (1.67) and (1.73) as

$$\frac{\partial}{\partial x} \lambda \frac{\partial p}{\partial x} = 0 \,, \tag{1.99}$$

$$v \frac{\partial S_w}{\partial x} + \phi \frac{\partial S_w}{\partial t} = 0 \,, \tag{1.100}$$

which are linear elliptic and hyperbolic (convection) equations with spatially varying coefficients, with

$$v = -\lambda \frac{\partial p}{\partial x} \,. \tag{1.101}$$

If an additional diffusion term is introduced in Eq. (1.100) we obtain a linear convection–diffusion equation

$$-D \frac{\partial^2 S_w}{\partial x^2} + v \frac{\partial S_w}{\partial x} + \phi \frac{\partial S_w}{\partial t} = 0 \,, \tag{1.102}$$

where D is the diffusion constant. In this case Eqs. (1.99) and (1.102) (or (1.100)) can be interpreted as describing the flow of two incompressible *miscible* fluids with identical properties such as water with two different colors (sometimes referred to as a *blue and red water* situation). Alternatively, the equations can be interpreted to describe the flow of *immiscible* fluids, in which case D represents the effect of dispersion due to geological heterogeneities.

1.4.7 Formation Volume Factors*

In our derivation we used equations of state (1.53) and (1.54) to relate pressure, temperature and densities of the reservoir fluids. These equations of state can also be expressed as relationships between pressure, temperature and volumes; e.g. for the oil we can write, instead of Eq. (1.53):

$$c_o \overset{\triangle}{=} -\frac{1}{V_o} \frac{\partial V_o}{\partial p_o}\bigg|_{T_R} \,. \tag{1.103}$$

In many practical reservoir engineering applications, the fluid densities and volumes at reservoir conditions are expressed in terms of those at *standard conditions*[12] with the aid of a *formation volume factor*. In particular gas and to a lesser extent oil change volume when flowing from the reservoir to surface. The oil formation volume factor B_o is defined as the ratio of a unit volume of oil at down hole conditions p and T, including dissolved gas, and the volume it occupies after it has been transferred to standard conditions p_{sc} and T_{sc}, during which journey gas has escaped from the oil:

$$B_o \triangleq \frac{V_o|_{p,T}}{V_o|_{p_{sc},T_{sc}}} . \tag{1.104}$$

An equivalent definition holds for the water formation volume factor, although usually very little gas dissolves in water. The gas formation volume factor is defined as the ratio of a unit volume of gas at down hole conditions, and the volume it occupies after it has been transferred to standard conditions, during which journey possibly some liquid drop-out may have occurred. Gas originally at reservoir conditions dramatically expands when the pressure approaches standard conditions, even if it loses some liquids, and the therefore B_g is typically much smaller than one. Oppositely, oil at reservoir conditions always contains a large amount of dissolved gas which escapes from the oil when transferred to standard conditions, and therefore B_o is always larger than one. Water hardly changes volume, compared to oil, and therefore B_w is always close to one. If we substitute Eq. (1.104) in Eq. (1.103), taking $p = p_o$, and $T = T_R$, it follows that

$$c_o \triangleq -\frac{1}{B_o}\frac{\partial B_o}{\partial p_o}\bigg|_{T_R}, \tag{1.105}$$

and similar expressions can be obtained for gas and water. Formation volume factors change with pressure and temperature, and therefore also the fluid compressibilities are functions of pressure and temperature. Even if we take the temperature constant at its reservoir value, the pressure dependence of the fluid compressibilities is often large enough to take them into account, thus introducing an additional nonlinearity in the reservoir simulation equations. This hold especially for gas reservoirs, and to a lesser extent for oil reservoirs that experience large pressure changes, e.g. during primary recovery. Determination of the pressure and temperature dependency of formation volume factors is usually done with the aid of laboratory experiments on fluid samples taken from exploration wells. In absence of samples, a large number of correlations available from literature can be used to estimate the pressure and temperature dependence of the reservoir fluids; see e.g.

---

[12] In the upstream oil industry standard conditions are usually defined as a pressure $p_{sc} = 100$ kPa (14.7 psi) and a temperature $T_{sc} = 15$ °C (60 °F), which can be considered as typical for atmospheric conditions in temperate climates. Oil at standard conditions is often referred to as *stock tank oil*.

Whitson and Brulé (2000). Most of the reservoir engineering literature traditionally uses pressure-dependent formation volume factors rather than pressure-dependent fluid compressibilities. For liquid flow, and as long as the pressure changes in the reservoir remain small compared to the total pressure, the nonlinearity of the oil compressibility remains small, while the water compressibility remains as good as constant, a situation that applies to all the examples that we use in this text. To keep the equations as simple as possible, we have therefore chosen not to use formation volume factors in our derivations, but to use (constant) compressibilities instead.

1.4.8 Finite-Difference Discretization

This section gives a brief overview of the semi-discretization of Eq. (1.58) with the aid of finite differences for 2-D flow. For details and for alternative discretization schemes we refer to the references mentioned in Sect. 1.1. Following the same approach as used for one-phase flow, we rewrite Eqs. (1.58) and (1.59) in scalar 2-D form, assuming isotropic permeability, pressure independence of the parameters, and absence of gravity forces[13]:

$$
-\frac{h}{\mu_w}\left[\frac{\partial}{\partial x}\left(kk_{rw}\frac{\partial p}{\partial x}\right) + \frac{\partial}{\partial y}\left(kk_{rw}\frac{\partial p}{\partial y}\right)\right] + h\left[\phi S_w(c_w + c_r)\frac{\partial p}{\partial t} + \phi\frac{\partial S_w}{\partial t}\right] - hq_w'''
$$
$$
= 0,
$$

$$(1.106)$$

$$
-\frac{h}{\mu_o}\left[\frac{\partial}{\partial x}\left(kk_{ro}\frac{\partial p}{\partial x}\right) + \frac{\partial}{\partial y}\left(kk_{ro}\frac{\partial p}{\partial y}\right)\right] + h\left[\phi(1 - S_w)(c_o + c_r)\frac{\partial p}{\partial t} - \phi\frac{\partial S_w}{\partial t}\right] - hq_o'''
$$
$$
= 0.
$$

$$(1.107)$$

The first term in Eq. (1.106) can be rewritten as

$$
\frac{h}{\mu_w}\frac{\partial}{\partial x}\left(kk_{rw}\frac{\partial p}{\partial x}\right) \approx \frac{h}{\mu_w}\frac{\Delta}{\Delta x}\left(kk_{rw}\frac{\Delta p}{\Delta x}\right)
$$
$$
= \frac{h}{\mu_w}\frac{(kk_{rw})_{i+\frac{1}{2},j}(p_{i+1,j} - p_{i,j}) - (kk_{rw})_{i-\frac{1}{2},j}(p_{i,j} - p_{i-1,j})}{(\Delta x)^2}, \quad (1.108)
$$

where the absolute permeabilities k are based on harmonic averages just as in the single-phase case; see Eq. (1.28). However, the relative permeabilities k_{rw} need to be determined through upstream weighting to obtain the correct convective behavior; see Aziz and Settari (1979), p. 153. This implies that

---

[13] To stay in line with the notation used in the single-phase flow case, we should have used ϕ_0, $\rho_{o,0}$ and $\rho_{w,0}$ to indicate the pressure-independence of these parameters, but we have dropped the subscripts 0 to simplify the notation.

$$(k_{rw})_{i+\frac{1}{2},j} \triangleq \begin{cases} (k_{rw})_{i,j} & \text{if} \quad p_{i,j} \geq p_{i+1,j} \\ (k_{rw})_{i+1,j} & \text{if} \quad p_{i,j} < p_{i+1,j} \end{cases}, \tag{1.109}$$

The second term in Eq. (1.106) can be rewritten in a similar fashion. Combining and reorganizing all terms results in

$$V\left[\phi S_w(c_w + c_r)\frac{\partial p}{\partial t} + \phi\frac{\partial S_w}{\partial t}\right]_{i,j} - (T_w)_{i-\frac{1}{2},j}p_{i-1,j} - (T_w)_{i,j-\frac{1}{2}}p_{i,j-1} +$$

$$\left[(T_w)_{i-\frac{1}{2},j} + (T_w)_{i,j-\frac{1}{2}} + (T_w)_{i,j+\frac{1}{2}} + (T_w)_{i+\frac{1}{2},j}\right]p_{i,j} - (T_w)_{i,j+\frac{1}{2}}p_{i,j+1} - (T_w)_{i+\frac{1}{2},j}p_{i+1,j} = V(q_w''')_{i,j}, \tag{1.110}$$

where the transmissibilities are now defined as

$$(T_w)_{i-\frac{1}{2},j} \triangleq \frac{\Delta y}{\Delta x}\frac{h}{\mu_w}(kk_{rw})_{i-\frac{1}{2},j}, \text{etc}. \tag{1.111}$$

A similar discretization can be obtained for Eq. (1.107):

$$V\left[\phi(1 - S_w)(c_o + c_r)\frac{\partial p}{\partial t} - \phi\frac{\partial S_w}{\partial t}\right]_{i,j} - (T_o)_{i-\frac{1}{2},j}p_{i-1,j} - (T_o)_{i,j-\frac{1}{2}}p_{i,j-1} +$$

$$\left[(T_o)_{i-\frac{1}{2},j} + (T_o)_{i,j-\frac{1}{2}} + (T_o)_{i,j+\frac{1}{2}} + (T_o)_{i+\frac{1}{2},j}\right]p_{i,j} - (T_o)_{i,j+\frac{1}{2}}p_{i,j+1} - (T_o)_{i+\frac{1}{2},j}p_{i+1,j} = V(q_o''')_{i,j}. \tag{1.112}$$

Equations (1.110) and (1.112) can be written in matrix form as

$$\begin{bmatrix} \mathbf{V}_{wp} & \mathbf{V}_{ws} \\ \mathbf{V}_{op} & \mathbf{V}_{os} \end{bmatrix}\begin{bmatrix} \dot{\mathbf{p}} \\ \dot{\mathbf{s}} \end{bmatrix} + \begin{bmatrix} \mathbf{T}_w & 0 \\ \mathbf{T}_o & 0 \end{bmatrix}\begin{bmatrix} \mathbf{p} \\ \mathbf{s} \end{bmatrix} = \begin{bmatrix} \mathbf{q}_w \\ \mathbf{q}_o \end{bmatrix}, \tag{1.113}$$

where the vectors $\mathbf{p}$ and $\mathbf{s}$ contain pressures and water saturations:

$$\mathbf{p}^T \triangleq [p_{i,j-1} \quad \cdots \quad p_{i-1,j} \quad p_{i,j} \quad p_{i+1,j} \quad \cdots \quad p_{i,j+1}], \tag{1.114}$$

$$\mathbf{s}^T \triangleq \left[(S_w)_{i,j-1} \quad \cdots \quad (S_w)_{i-1,j} \quad (S_w)_{i,j} \quad (S_w)_{i+1,j} \quad \cdots \quad (S_w)_{i,j+1}\right], \tag{1.115}$$

and where the sub-matrices $\mathbf{V}_{wp}$, $\mathbf{V}_{ws}$, $\mathbf{V}_{op}$ and $\mathbf{V}_{os}$ contain accumulation terms[14]

$$\mathbf{V}_{wp} \triangleq V(c_w + c_r)\left[0 \quad \cdots \quad 0 \quad \phi_{i,j} \times (S_w)_{i,j} \quad 0 \quad \cdots \quad 0\right], \tag{1.116}$$

$$\mathbf{V}_{ws} \triangleq V\left[0 \quad \cdots \quad 0 \quad \phi_{i,j} \quad 0 \quad \cdots \quad 0\right], \tag{1.117}$$

$$\mathbf{V}_{op} \triangleq V(c_o + c_r)\left[0 \quad \cdots \quad 0 \quad \phi_{i,j} \times (1 - S_w)_{i,j} \quad 0 \quad \cdots \quad 0\right], \tag{1.118}$$

$$\mathbf{V}_{os} \triangleq -V\left[0 \quad \cdots \quad 0 \quad \phi_{i,j} \quad 0 \quad \cdots \quad 0\right], \tag{1.119}$$

---

[14] Here the sub-matrices are displayed as vectors, which form, however, building blocks for matrices when the equations for multiple grid blocks are combined.

the sub-matrices $\mathbf{T}_w$ and $\mathbf{T}_o$ contain transmissibility terms

$$\mathbf{T}_w \triangleq \left[-(T_w)_{i,j-\frac{1}{2}} \quad \cdots \quad -(T_w)_{i-\frac{1}{2},j} \quad \left((T_w)_{i,j-\frac{1}{2}} + (T_w)_{i-\frac{1}{2},j} + (T_w)_{i+\frac{1}{2},j} + (T_w)_{i,j+\frac{1}{2}} \right) \quad -(T_w)_{i+\frac{1}{2},j} \quad \cdots \quad -(T_w)_{i,j+\frac{1}{2}} \right],$$
(1.120)

$$\mathbf{T}_o \triangleq \left[-(T_o)_{i,j-\frac{1}{2}} \quad \cdots \quad -(T_o)_{i-\frac{1}{2},j} \quad \left((T_o)_{i,j-\frac{1}{2}} + (T_o)_{i-\frac{1}{2},j} + (T_o)_{i+\frac{1}{2},j} + (T_o)_{i,j+\frac{1}{2}} \right) \quad -(T_o)_{i+\frac{1}{2},j} \quad \cdots \quad -(T_o)_{i,j+\frac{1}{2}} \right],$$
(1.121)

and vectors $\mathbf{q}_w$ and $\mathbf{q}_o$ contain the flow rates (source terms) with elements expressed in m³/s:

$$\mathbf{q}_w^T \triangleq \left[\cdots \quad (q_w)_{i,j} \quad \cdots \right],$$
(1.122)

$$\mathbf{q}_o^T \triangleq \left[\cdots \quad (q_o)_{i,j} \quad \cdots \right].$$
(1.123)

1.4.9 Example 3: Inverted Five-Spot

Example 3 concerns a square reservoir with heterogeneous permeability and porosity fields as depicted in Fig. 1.8, modeled with $21 \times 21 = 441$ grid blocks. It can be seen that the permeability displays a marked streak running from the South-West (bottom left) corner to just below the North-East (top right) corner, and that the porosity is mildly correlated with the permeability. The other relevant parameters have been listed in Table 1.4. As in the earlier examples, gravity and capillary forces are neglected.

Figure 1.9 illustrates the sparse structure of the secant matrices for Example 3. The top-left figure corresponds to the accumulation matrix $\mathbf{V}$ which has a perfect diagonal structure in each of its four quadrants. The top-right figure corresponds to the transmissibility matrix $\mathbf{T}$ which has an almost penta-diagonal structure (tri-diagonal with 'holes' and two side bands) in the two left quadrants. The two

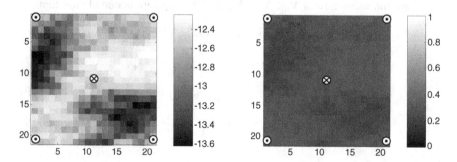

Fig. 1.8 Well configuration, and permeability and porosity fields for Example 3. The reservoir is produced with a central injector and five producers in the corners

Table 1.4 Reservoir and fluid properties for Example 3

Symbol	Variable	Value	SI units	Value	Field units
h	Grid-block height	2	m	6.56	ft
$\Delta x, \Delta y$	Grid-block length/width	33.33	m	109.36	ft
μ_o	Oil dynamic viscosity	5.0×10^{-4}	Pa s	0.5	cP
μ_w	Water dynamic viscosity	1.0×10^{-3}	Pa s	1.0	cP
c_t	Total compressibility	3.0×10^{-9}	Pa^{-1}	2.1×10^{-5}	psi^{-1}
$\breve{p}_R$	Initial reservoir pressure	30×10^6	Pa	4351.1	psi
r_{well}	Well-bore radius	0.114	m	4.50	in
k_{ro}^0	End-point relative permeability, oil			0.9	–
k_{rw}^0	End-point relative permeability, water			0.6	–
n_o	Corey exponent, oil			2.0	–
n_w	Corey exponent, water			2.0	–
S_{or}	Residual-oil saturation			0.2	–
S_{wc}	Connate-water saturation			0.2	–

right quadrants are completely filled with zeros because we have neglected capillary forces. The bottom-left and bottom-right figures display details of **T** and illustrate the side bands and the 'holes' in the tri-diagonals at every 21st row which are typical for a regular numbering scheme. In this case the 441 grid blocks have been numbered row-wise from top-left to bottom-right. A grid block i in the center of the grid, i.e. not at an edge or at a corner, is connected to its Western and Eastern neighbors with numbers $i - 1$ and $i + 1$ respectively, which results in the tri-diagonals, and to its Northern and Southern neighbors $i - 21$ and $i + 21$ respectively which results in the penta-diagonal side bands. Grid blocks at an edge are missing one of the connections which results in irregularities in the structure in the form of 'holes' in the tri-diagonals.

1.4.10 Sources of Nonlinearity

Although we use a matrix–vector notation, which suggests a linear system of equations, Eq. (1.113) are nonlinear because the coefficients of sub-matrices $\mathbf{V}_{wp}$, $\mathbf{V}_{op}$ $\mathbf{T}_w$ and $\mathbf{T}_o$ are functions of S_w. In particular, the coefficients of the transmissibility matrices $\mathbf{T}_w$ and $\mathbf{T}_o$ contain saturation-dependent relative permeabilities, for which we use the Corey model given by Eqs. (1.74) to (1.76) Actually, the coefficients of the transmissibility matrices are also functions of p because of the upstream weighting of the relative permeabilities. That is, if the pressures in two adjacent grid blocks change slightly, but such that the flow through the grid-block boundary changes direction, the upstream relative permeability and therefore the transmissibility may change strongly. This nonlinear effect may cause problems during the iterative solution of the system equations during implicit time integration, if the flow direction keeps changing during subsequent iterations. However,

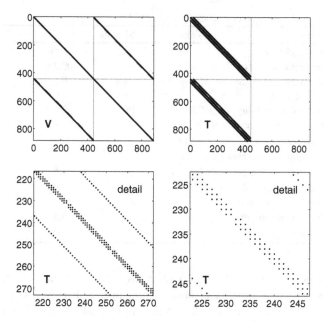

Fig. 1.9 System matrices for Example 3

because it is a discontinuous nonlinearity, we cannot differentiate the transmissibilities with respect to p, and we cannot take it into account during linearization of the equations as required for e.g. implicit integration with Newton–Raphson iteration. Another source of nonlinearity are the source terms $\mathbf{q}_o$ and $\mathbf{q}_w$ in Eq. (1.113) which cannot always be prescribed directly. In the case of a water injection well, the oil flow rates are equal to zero, and it is possible to prescribe the water injection rates. In a production well, however, the proportions of oil and water in the total flow rate q_t depend on the fractional flows f_o and f_w, i.e. on the relative magnitude of the oil and water mobilities around the well according to

$$q_o = f_o q_t = \frac{\lambda_o}{\lambda_o + \lambda_w} q_t, \tag{1.124}$$

$$q_w = f_w q_t = \frac{\lambda_w}{\lambda_o + \lambda_w} q_t, \tag{1.125}$$

where the saturation-dependent mobilities λ_o and λ_w are given by Eqs. (1.62) and (1.63). Therefore we need to specify

$$\underbrace{\begin{bmatrix} \mathbf{q}_w \\ \mathbf{q}_o \end{bmatrix}}_{\mathbf{q}} = \underbrace{\begin{bmatrix} \mathbf{F}_w(\mathbf{s}) \\ \mathbf{F}_o(\mathbf{s}) \end{bmatrix}}_{\mathbf{F}} \mathbf{q}_t, \tag{1.126}$$

where $\mathbf{F}_o$ and $\mathbf{F}_w$ are diagonal matrices of which the non-zero entries contain fractional flows:

$$\mathbf{F}_w \triangleq \begin{bmatrix} 0 & \cdots & 0 & (f_w)_{i,j} & 0 & \cdots & 0 \end{bmatrix}; \qquad (1.127)$$

$$\mathbf{F}_o \triangleq \begin{bmatrix} 0 & \cdots & 0 & (f_o)_{i,j} & 0 & \cdots & 0 \end{bmatrix}. \qquad (1.128)$$

To emphasize the nonlinearities, Eq. (1.113) may therefore be rewritten as[15]

$$\begin{bmatrix} \mathbf{V}_{wp}(\mathbf{s}) & \mathbf{V}_{ws} \\ \mathbf{V}_{op}(\mathbf{s}) & \mathbf{V}_{os} \end{bmatrix} \begin{bmatrix} \dot{\mathbf{p}} \\ \dot{\mathbf{s}} \end{bmatrix} + \begin{bmatrix} \mathbf{T}_w(\mathbf{s}) & \mathbf{0} \\ \mathbf{T}_o(\mathbf{s}) & \mathbf{0} \end{bmatrix} \begin{bmatrix} \mathbf{p} \\ \mathbf{s} \end{bmatrix} = \begin{bmatrix} \mathbf{F}_w(\mathbf{s}) \\ \mathbf{F}_o(\mathbf{s}) \end{bmatrix} \mathbf{q}_t. \qquad (1.129)$$

In an injection well we have $\mathbf{q}_t = \mathbf{q}_w$, and we expect that soon after injection has started the fractional flows for water and oil close to the well will approach one and zero respectively. However, before injection starts, the initial condition for the saturation is usually equal to the connate-water saturation, which means that the fractional flows for water and oil are zero and one respectively, which implies that it is impossible to ever inject water. This paradox is usually circumvented by simply specifying a fractional flow equal to one for every injection well.

1.4.11 Incompressible Flow

In the special case that the fluid and rock compressibilities are so small that they may be neglected, it follows from Eqs. (1.116) and (1.118) that $\mathbf{V}_{wp} = \mathbf{V}_{op} = \mathbf{0}$. In that case we can rewrite Eq. (1.129) as

$$\begin{bmatrix} \mathbf{0} & \mathbf{V}_{ws} \\ \mathbf{0} & \mathbf{V}_{os} \end{bmatrix} \begin{bmatrix} \dot{\mathbf{p}} \\ \dot{\mathbf{s}} \end{bmatrix} + \begin{bmatrix} \mathbf{T}_w(\mathbf{s}) & \mathbf{0} \\ \mathbf{T}_o(\mathbf{s}) & \mathbf{0} \end{bmatrix} \begin{bmatrix} \mathbf{p} \\ \mathbf{s} \end{bmatrix} = \begin{bmatrix} \mathbf{F}_w(\mathbf{s}) \\ \mathbf{F}_o(\mathbf{s}) \end{bmatrix} \mathbf{q}_t, \qquad (1.130)$$

which may also be expressed as:

$$\mathbf{V}_{ws}\dot{\mathbf{s}} + \mathbf{T}_w(\mathbf{s})\mathbf{p} = \mathbf{F}_w(\mathbf{s})\mathbf{q}_t, \qquad (1.131)$$

$$\mathbf{V}_{os}\dot{\mathbf{s}} + \mathbf{T}_o(\mathbf{s})\mathbf{p} = \mathbf{F}_o(\mathbf{s})\mathbf{q}_t. \qquad (1.132)$$

Because $\mathbf{V}_{ws} = -\mathbf{V}_{os}$ we can add the two equations to obtain the pressure equation for incompressible flow

$$\mathbf{T}_t(\mathbf{s})\mathbf{p} = \mathbf{q}_t, \qquad (1.133)$$

where we used the equality $\mathbf{F}_w + \mathbf{F}_o = \mathbf{I}$, and where $\mathbf{T}_t = \mathbf{T}_w + \mathbf{T}_o$ is the total transmissibility matrix, which is still a function of saturation. Note that the

---

[15] As discussed before, we disregard the dependency of the transmissibility terms on pressure.

pressure equation is no longer a differential equation but has degenerated to an algebraic equation. The physical background is that the vanishing of compressibilities means that there is no longer a possibility to store energy in the system. Just as in the case of single-phase flow, discussed in Sect. 1.3.4, it appears as if Eq. (1.133) can simply be solved for the constant pressure **p**. However, the total transmissibility matrix $\mathbf{T}_t$ is singular which implies that we cannot directly solve the equation. Just as in the single-phase case, the singularity can be removed through prescribing the pressure in at least one of the grid blocks, or through the use of a well model, in which case we may indeed solve for **p**. Thereafter, one of the two Eqs. (1.131) or (1.132) can be used to compute the water saturations.

1.4.12 Fluid Velocities*

1.4.12.1 Total Velocity*

Darcy's law, which specifies an empirical relationship between pressure gradients and fluid velocities, is at the heart of the description of flow through porous media. However, in our formulation of the flow equations in the previous sections we rapidly lost the fluid velocities as variables, through substitution in the mass-balance equations; see Eqs. (1.4), (1.7) and (1.10) for the single-phase flow case, and Eqs. (1.45)–(1.50) for the two-phase case. To recover the fluid velocities, after solving the flow equations, we have to revert to Darcy's law. Often we are interested in the *total velocity* which is the sum of the phase velocities. In the case of two-phase oil–water flow, Darcy's law for the total velocity is obtained by adding Eqs. (1.47) and (1.48):

$$\vec{\mathbf{v}}_t = -\frac{k_{rw}}{\mu_w}\vec{\mathbf{K}}(\nabla p_w - \rho_w g \nabla d) - \frac{k_{ro}}{\mu_o}\vec{\mathbf{K}}(\nabla p_o - \rho_o g \nabla d). \qquad (1.134)$$

If, as before, we neglect gravity and capillary forces, assume isotropic permeability, and express the equations in scalar form we obtain

$$v_{t,x} = -k\left(\frac{k_{rw}}{\mu_w} + \frac{k_{ro}}{\mu_o}\right)\frac{\partial p}{\partial x}, \qquad (1.135)$$

$$v_{t,y} = -k\left(\frac{k_{rw}}{\mu_w} + \frac{k_{ro}}{\mu_o}\right)\frac{\partial p}{\partial y}. \qquad (1.136)$$

Starting from these equations we can now use a finite-difference discretization to find numerical approximations of the velocity components:

$$v_{t,x} \approx (v_t)_{i+\frac{1}{2},j} \overset{\Delta}{=} -\left(\frac{kk_{rw}}{\mu_w} + \frac{kk_{ro}}{\mu_o}\right)_{i+\frac{1}{2},j}\frac{p_{i+1,j} - p_{i,j}}{\Delta x}, \qquad (1.137)$$

$$v_{t,y} \approx (v_t)_{i,j+\frac{1}{2}} \triangleq - \left(\frac{kk_{rw}}{\mu_w} + \frac{kk_{ro}}{\mu_o} \right)_{i,j+\frac{1}{2}} \frac{p_{i,j+1} - p_{i,j}}{\Delta y} . \qquad (1.138)$$

In these discretized equations the pressures are taken in the grid-block centers whereas the velocities and the (averaged) parameters are taken at the grid-block boundaries. Note that Eqs. (1.137) and (1.138) represent the velocities at the left and bottom boundaries of a grid block. Similar expressions can be obtained for the right and top boundaries.

1.4.12.2 Velocities at Grid-Block Boundaries*

In the case of tracing streamlines, as will be discussed in Sect. 2.3.6, we need the velocities at the grid-block boundaries which can be obtained from Eqs. (1.137) and (1.138). Alternatively, we can use the following, slightly different, approach that directly exploits the connectivity structure of the grid blocks. Consider a block-centered finite-difference reservoir model with n_{gb} grid blocks and n_{con} grid-block connectivities. If we represent the grid-block pressures by an $n_{gb} \times 1$ vector $\mathbf{p}$, we can define a linear transformation

$$p = \mathbf{L}_{pp}\mathbf{p} , \qquad (1.139)$$

where p is an $n_{con} \times 1$ vector of pressure differences between the grid-block centers,[16] and $\mathbf{L}_{pp}$ is an $n_{con} \times n_{gb}$ selection matrix with entries -1, 0 and 1 in the appropriate places.[17] E.g. for our 6-grid-block Example 1 we find that $\mathbf{L}_{pp}$ is a 7×6 matrix given by

$$\mathbf{L}_{pp} = \begin{bmatrix} -1 & 1 & 0 & 0 & 0 & 0 \\ -1 & 0 & 0 & 1 & 0 & 0 \\ 0 & -1 & 1 & 0 & 0 & 0 \\ 0 & -1 & 0 & 0 & 1 & 0 \\ 0 & 0 & -1 & 0 & 0 & 1 \\ 0 & 0 & 0 & -1 & 1 & 0 \\ 0 & 0 & 0 & 0 & -1 & 1 \end{bmatrix} . \qquad (1.140)$$

In passing we note that we can almost never invert Eq. (1.139) to reconstruct the pressure vector $\mathbf{p}$ from the pressure difference vector p, because $\mathbf{L}_{pp}$ is nearly

---

[16] We use a **bold-face italics** font to represent vectors with properties at the grid-blocks boundaries, whereas the conventional **bold-face** font is used to represent vectors with properties at the grid-blocks centers. In particular we use p and $\mathbf{p}$ for pressures, λ and $\boldsymbol{\lambda}$ for mobilities, v and $\mathbf{v}$ for Darcy velocities, and $\bar{v}$ and $\bar{\mathbf{v}}$ for interstitial velocities.

[17] This is an example of an *incidence matrix* as used in the analysis of e.g. electrical or mechanical networks, to define the pattern of nodes and edges. Other names used in the literature are *topology matrix* or *connectivity matrix*. Note that we use the term connectivity in a slightly different sense; see p. 69.

always rectangular with $n_{con} > n_{gb}$. This illustrates that knowledge of just the inter-grid-block pressure *differences* does not give us enough information to compute the absolute pressures in the grid blocks. The $n_{con} \times 1$ vector v_t of total Darcy velocities over the grid-block boundaries can now be written as a function of p according to

$$v_t = -diag(\lambda)diag(\gamma)\, p\,. \tag{1.141}$$

Here λ is an $n_{con} \times 1$ vector of averaged mobilities,

$$\lambda_i \triangleq \left(\frac{k_i k_{ro,i}}{\mu_o} + \frac{k_i k_{rw,i}}{\mu_w} \right), \quad i = 1, \ldots, n_{con}\,, \tag{1.142}$$

where the absolute permeabilities k_i are harmonic averages of the adjacent grid blocks while the relative permeabilities $k_{ro,i}$ and $k_{rw,i}$ are usually upstream-weighted. The vector γ is another $n_{con} \times 1$ vector with elements γ_i that are geometric factors depending on the grid-block properties. For example in the case of flow in the x direction γ is simply equal to $1/\Delta x$. Combining Eqs. (1.139) and (1.141) we can write

$$v_t = \mathbf{S} p\,, \tag{1.143}$$

where the $n_{con} \times n_{gb}$ matrix $\mathbf{S}$ is given by

$$\mathbf{S} = -diag(\lambda)diag(\gamma)\, \mathbf{L}_{pp}\,. \tag{1.144}$$

Equation (1.144) can be used to compute the inter-grid-block velocities for streamline tracking as described in Sect. 2.3.6.

References

Aarnes JE, Gimse T, Lie KA (2007) An introduction to the numerics of flow in porous media using Matlab. In: Hasle G, Lie KA, Quack E (eds) *Geometric modeling, numerical simulation, and optimization; applied mathematics at SINTEF*. Springer, Berlin

Aziz K, Settari A (1979) Petroleum reservoir simulation. Applied Science Publishers, London

Bear J (1972) Dynamics of fluids in porous media. Elsevier, New York. Reprinted in 1988 by Dover, New York

Buckley SE, Leverett MC (1942) Mechanisms of fluid displacement in sands. Pet Trans AIME 146:107–116

Chen Z, Huan G, Ma Y (2006) Computational methods for multiphase flows in porous media. SIAM, Philadelphia

Ding Y, Lemonnier PA, Estebenet T, Magras J-F (2000) Control-volume method for simulation in the well vicinity for arbitrary well configurations. SPE J 5(1):118–125. doi:10.2118/62169-PA

Ewing RE (1983) Problems arising in the modeling of processes for hydrocarbon recovery. In: Ewing RE (ed) The mathematics of reservoir simulation. SIAM, Philadelphia

Fanchi JR (2006) Principles of applied reservoir simulation, 3rd edn. Gulf Professional Publishing, Burlington

Gerritsen MG, Durlofsky LJ (2005) Modeling fluid flow in oil reservoirs. Annu Rev Fluid Mech 37:211–238. doi:10.1146/annurev.fluid.37.061903.175748

Helmig R (1997) Multiphase flow and transport processes in the subsurface. Springer, Berlin

Jansen JD, Bosgra OH, Van den Hof PMJ (2008) Model-based control of multiphase flow in subsurface oil reservoirs. J Process Control 18:846–855. doi:10.1016/j.jprocont.2008.06.011

Lake LW (1989) Enhanced oil recovery. Prentice Hall, Upper Saddle River

Mattax CC, Dalton RL (1990) Reservoir simulation. SPE Monograph Series 13, SPE, Richardson

Oliver DS, Reynolds AC, Liu N (2008) Inverse theory for petroleum reservoir characterization and history matching. Cambridge University Press, Cambridge

Patankar SV (1980) Numerical heat transfer and fluid flow. Series in computational methods and in mechanics and thermal sciences, Taylor and Francis

Peaceman DW (1977) Fundamentals of numerical reservoir simulation. Elsevier, Amsterdam

Peaceman DW (1978) Interpretation of well-block pressures in numerical reservoir simulation. SPE J 18(3):183–194. doi:10.2118/6893-PA

Peaceman DW (1983) Interpretation of well-block pressures in numerical reservoir simulation with nonsquare grid blocks and anisotropic permeability. SPE J 23(3):531–543. doi:10.2118/10528-PA

Russel TF, Wheeler MF (1983) Finite-element and finite-difference methods for continuous flows in porous media. In: Ewing RE (ed) The mathematics of reservoir simulation. SIAM, Philadelphia

Welge HJ (1952) A simplified method for computing oil recovery by gas or water drive. Pet Trans AIME 195:91–98

Whitson CH, Brulé MR (2000) Phase behavior. SPE Monograph Series 20, SPE, Richardson

Chapter 2
System Models

Abstract This chapter develops representations in state-space notation of the porous-media flow equations derived in Chap. 1. For single-phase flow, the states are grid-block pressures, and for two-phase flow they are grid-block pressures and saturations. The inputs are typically bottom-hole pressures or total well flow rates, the outputs are typically bottom-hole pressures in those wells were the flow rates were prescribed, and phase rates in those wells were the bottom-hole pressures were prescribed. The use of matrix partitioning to describe the different types of inputs leads to a description in terms of nonlinear ordinary-differential and algebraic equations with (state-dependent) system, input, output and direct-throughput matrices. Other topics include generalized state-space representations, linearization, elimination of prescribed pressures, the tracing of stream lines, lift tables, computational aspects, and the derivation of an energy balance for porous-media flow.

2.1 System Equations

2.1.1 Partial-Differential Equations

To describe the physics of fluid flow in a porous medium we generally use *partial-differential equations* (PDEs). Typically the *independent* variables are time, t, and spatial coordinates x, y and z. Furthermore, in reservoir engineering we normally encounter only first-order derivatives in time, but higher-order (typically second-order) derivatives in space. Indicating the, arbitrary, *dependent* variable with a fat dot, $\bullet$, the equations can be represented in general form as[1]

---

[1] The dependent variables follow from the physics of the problem. In case of multi-phase flow through porous media they are typically pressures, component masses or phase saturations; see Chap. 1. Here we use only a single dependent variable, but in general multiple dependent variables will occur, in which case multiple differential equations are required to describe the problem.

J. D. Jansen, *A Systems Description of Flow Through Porous Media*,
SpringerBriefs in Earth Sciences, DOI: 10.1007/978-3-319-00260-6_2,
© The Author(s) 2013

$$\varepsilon(t,x,y,z,\bullet) \times \frac{\partial(\bullet)}{\partial t} = \varphi(t,x,y,z,\bullet) \times L(\bullet) + \psi(t,x,y,z,\bullet), \qquad (2.1)$$

where ε and φ are parameters that may be functions of time and space, L is a spatial *differential operator* and ψ is the *source term*. Note that ε, φ and ψ may be functions of the dependent variable, in which case the equation is nonlinear. The left-hand term in Eq. (2.1) is known as the *accumulation term*, the first term at the right-hand side as the *transport term*. A specific example of the general PDE (2.1) is the mass conservation Eq. 1.4 in Sect. 1.3.1 in which case the spatial difference operator L is the divergence $\nabla(\bullet) \triangleq \partial(\bullet)/\partial x + \partial(\bullet)/\partial y + \partial(\bullet)/\partial z$. In addition to the PDE (2.1), we need to specify the spatial domain Ω and the time domain T on which it is valid. At the boundary Γ of Ω we need to specify boundary conditions, and at a specific point in time an initial condition, to complete the problem formulation. A PDE describes the evolution of the dependent variables in time and space in a continuous fashion, and the solution is therefore specified in an infinitely large number of points. Closed-form solutions of PDEs, i.e. to 'infinite dimensional problems', are generally restricted to simple domains and parameters that are spatially homogeneous. For more realistic geometries we need to solve the equations numerically, which requires some form of *discretization* of the equations, because digital computers can only deal with finite dimensional problems.

2.1.2 Ordinary-Differential Equations

An often followed approach is to first perform a spatial discretization of the PDEs, and only perform the time discretization at a later stage. The initial *semi-discretization* of the equations, i.e. the discretization in space, can be done using the method of finite differences, finite volumes or finite elements. An example of a finite-difference discretization as applied to porous-media flow has been worked out in Chap. 1. All of the discretization methods result in systems of *ordinary-differential equations* (ODEs) which can typically be represented as

$$\begin{cases} \hat{e}_1\left(\bullet_1, \dfrac{d(\bullet_1)}{dt}\right) = \hat{f}_1(t, \bullet_1, \bullet_2, \ldots, \bullet_n, \psi_1), \\[2mm] \hat{e}_2\left(\bullet_2, \dfrac{d(\bullet_2)}{dt}\right) = \hat{f}_2(t, \bullet_1, \bullet_2, \ldots, \bullet_n, \psi_2), \\[2mm] \qquad\qquad\vdots \\[2mm] \hat{e}_n\left(\bullet_n, \dfrac{d(\bullet_n)}{dt}\right) = \hat{f}_n(t, \bullet_1, \bullet_2, \ldots, \bullet_n, \psi_n), \end{cases} \qquad (2.2)$$

where the *continuous* dependent variable $\bullet$ and the continuous source term ψ of Eq. (2.1) are now represented with a finite number of *discrete* values $\bullet_i$ and ψ_i, corresponding to discrete points in space, and where $\hat{e}_i$ and $\hat{f}_i$ are, in the general case, nonlinear functions.[2] Normally the functions $\hat{e}_i$ are linear in the derivatives $d(\bullet_i)/dt$, which makes it possible to transform the system of Eq. (2.2) such that the derivatives in the left-hand side terms are isolated, leading to:

$$
\begin{cases}
\dfrac{d(\bullet_1)}{dt} = f_1(t, \bullet_1, \bullet_2, \ldots, \bullet_n, \psi_1), \\[2mm]
\dfrac{d(\bullet_2)}{dt} = f_2(t, \bullet_1, \bullet_2, \ldots, \bullet_n, \psi_2), \\[2mm]
\quad\vdots \\[2mm]
\dfrac{d(\bullet_n)}{dt} = f_n(t, \bullet_1, \bullet_2, \ldots, \bullet_n, \psi_n),
\end{cases}
\tag{2.3}
$$

where the functions f_i are different from the functions $\hat{f}_i$ in Eq. (2.2). Note that Eqs. (2.2) and (2.3) are both coupled systems of ODEs because each of the dependent variables $\bullet_i$ is present in more than one equation. If the functions f_i are linear in the dependent variables, a further simplification is possible that decouples the equations leading to

$$
\frac{d(\tilde{\bullet}_i)}{dt} = \tilde{f}_i\left(t, \tilde{\bullet}_i, \tilde{\psi}_i\right), \quad i = 1, 2, \ldots, n
\tag{2.4}
$$

where the transformed dependent variables $\tilde{\bullet}_i$ are linear combinations of the original variables $\bullet_i$, and $\tilde{f}_i$ are functions again. This decoupling procedure will be addressed in more detail in Sect. 3.1.2 below. In reservoir simulation the functions $\hat{e}_i$ and $\hat{f}_i$ are typically linear in the derivatives and nonlinear in the dependent variables, which at first sight implies that Eq. (2.3) is the most relevant representation. Moreover, system-theoretical results are usually derived using equations of this particular form. However, for large scale computations it is more efficient to use representation (2.2), and in this text we will therefore make use of both formulations.

2.1.3 State-Space Representation

2.1.3.1 State Equations

A more concise form of Eq. (2.3) can be obtained through the use of vector notation. We introduce the vectors $\mathbf{x} = [x_1 \quad x_2 \quad \ldots \quad x_n]^T$ and $\mathbf{u} = [u_1 \quad u_2 \quad \ldots \quad u_m]^T$ to represent the discrete values of the dependent variables, instead of $\bullet_i$ and ψ_i

---

[2] Typically, most of the values $\bullet_i$ are equal to zero in a single equation. For example in the case of one-dimensional single-phase flow modeled with first-order finite differences, the only three non-zero values $\bullet_i$ in the ith equation in a system of n equations with $1 < i < n$ are given by: $\hat{e}_i(\bullet_i, d(\bullet_i)/dt) = \hat{f}_i(\bullet_{i-1}, \bullet_i, \bullet_{i+1}, \psi_i)$.

which we used until now. The reason to use $\mathbf{x}$ and $\mathbf{u}$ is to adhere to the notation convention in the systems-and-control literature. Note that $x_1, x_2, \ldots x_n$ do *not* represent spatial coordinates. Also note that we have indicated that the source term $\mathbf{u}$ has m elements instead of n. This anticipates a situation where many of the source terms are equal to zero, such that $m \ll n$, in which case it may be computationally advantageous to use a shorter vector $\mathbf{u}$. Equations (2.3) can now be written as

$$\dot{\mathbf{x}}(t) = \mathbf{f}(t, \mathbf{u}(t), \dot{\mathbf{x}}(t)) \,, \tag{2.5}$$

where $\mathbf{f}$ is a nonlinear vector function of $\mathbf{x}$, $\mathbf{u}$ and t, and where we have emphasized the dependence of $\mathbf{x}$ and $\mathbf{u}$ on t by writing $\mathbf{x}(t)$ and $\mathbf{u}(t)$. In reservoir simulation, the equations are usually nonlinear but with coefficients that do not depend on time directly, such that we can write

$$\dot{\mathbf{x}}(t) = \mathbf{f}(\mathbf{u}(t), \mathbf{x}(t)) \,. \tag{2.6}$$

In the special case that $\mathbf{f}$ is a linear function of $\mathbf{x}$ and $\mathbf{u}$, we can use a vector-matrix notation and write Eq. (2.6) as a *linear time-varying* (LTV) vector differential equation

$$\dot{\mathbf{x}}(t) = \mathbf{A}(t)\mathbf{x}(t) + \mathbf{B}(t)\mathbf{u}(t) \,, \tag{2.7}$$

where the coefficients of the $n \times n$ matrix $\mathbf{A}$ and the $n \times m$ matrix $\mathbf{B}$ may still be functions of t. The matrices $\mathbf{A}$ and $\mathbf{B}$ are usually referred to as the *system matrix* and the *input matrix* respectively.[3] In the case that $\mathbf{A}$ and $\mathbf{B}$ are independent of t, we obtain a *linear time-invariant* (LTI) equation given by

$$\dot{\mathbf{x}}(t) = \mathbf{A}\mathbf{x}(t) + \mathbf{B}\mathbf{u}(t) \,. \tag{2.8}$$

From now on we will mostly not explicitly indicate the dependence on time of the variables, and we will write, e.g., $\dot{\mathbf{x}} = \mathbf{f}(\mathbf{u}, \mathbf{x})$ instead of $\dot{\mathbf{x}}(t) = \mathbf{f}(\mathbf{u}(t), \mathbf{x}(t))$. First-order systems of equations such as (2.5), (2.6), (2.7) and (2.8) are referred to as *state equations* in the systems-and-control literature. In this representation, the elements of vector $\mathbf{x}$ are the state variables which completely define the dynamic state of the system. A continuous sequence of values of $\mathbf{x}$ over a certain time interval is often referred to as a *trajectory* in state space. In reservoir engineering applications it is sometimes preferred to start from Eqs. (2.2) rather than from Eq. (2.3), even if the functions e_i are linear. We will refer to equations of the type of Eq. (2.2) as *generalized state equations*.[4] In LTI form they can be written as

$$\hat{\mathbf{E}}\dot{\mathbf{x}} = \hat{\mathbf{A}}\mathbf{x} + \hat{\mathbf{B}}\mathbf{u} \,. \tag{2.9}$$

---

[3] An alternative name for the input matrix is *distribution matrix* because it distributes the inputs $\mathbf{u}$ over the states $\mathbf{x}$.

[4] Sometimes this form of generalized state equations is referred to as a *descriptor system*.

2.1.3.2 Output Equations

In addition to Eqs. (2.5) to (2.9), which define the relationship between the input vector **u** and the state vector **x** of a dynamic system, we can also define a relationship between an output vector **y** and the state **x**. Moreover, the output may to some extent also depend directly on the input **u**, such that we can write

$$\mathbf{y} = \mathbf{h}(\mathbf{u}, \mathbf{x}), \tag{2.10}$$

for the nonlinear case or

$$\mathbf{y} = \mathbf{C}\mathbf{x} + \mathbf{D}\mathbf{u}, \tag{2.11}$$

for the linear case, where **C** is known as the *output matrix* and **D** as the *direct-throughput matrix*. If the output vector **y** has p elements, the matrices **C** and **D** have dimensions $p \times n$ and $p \times m$, respectively.

2.1.3.3 Implicit Nonlinear Equations

In addition to the general nonlinear system functions (2.6) and (2.10) we will sometimes use even more general nonlinear functions

$$\mathbf{g}(\mathbf{u}, \mathbf{x}, \dot{\mathbf{x}}) = \mathbf{0}, \tag{2.12}$$

$$\mathbf{j}(\mathbf{u}, \mathbf{x}, \mathbf{y}) = \mathbf{0}, \tag{2.13}$$

where **g** and **j** are nonlinear vector-valued functions.[5] Note that the explicit Eqs. (2.6) and (2.10) can always simply be expressed in the implicit form of Eqs. (2.12) and (2.13), i.e.

$$\mathbf{g}(\mathbf{u}, \mathbf{x}, \dot{\mathbf{x}}) = \dot{\mathbf{x}} - \mathbf{f}(\mathbf{u}, \mathbf{x}), \tag{2.14}$$

$$\mathbf{j}(\mathbf{u}, \mathbf{x}, \mathbf{y}) = \mathbf{y} - \mathbf{h}(\mathbf{u}, \mathbf{x}). \tag{2.15}$$

The reverse is not always true, i.e. it may not be possible to derive an explicit expression **f** for a given implicit representation **g**. However, usually the implicit representation may be solved numerically for $\dot{\mathbf{x}}$, typically using some form of time discretization and an iterative algorithm. In that case we can still conceptually write the nonlinear equations in their explicit forms (2.6) and (2.10) which is often preferred for analysis purposes. In most cases the functions **f**, **g**, **h** and **j** are to be interpreted as numerical operations, e.g. **g** could represent a complete reservoir simulator. Detailed examples of the state variable description of reservoir systems will be discussed below.

---

[5] System equations expressed as $\mathbf{g}(\ldots) = \mathbf{0}$ are sometimes referred to as equations in *residual* form.

2.1.3.4 Error Terms

In systems-and-control applications it is customary to introduce *error terms* to account for the fact that a system description is only an approximation of reality. For example we can write

$$\dot{\mathbf{x}} = \mathbf{Ax} + \mathbf{Bu} + \boldsymbol{\varepsilon}\,, \tag{2.16}$$

$$\mathbf{y} = \mathbf{Cx} + \mathbf{Du} + \boldsymbol{\eta}\,, \tag{2.17}$$

for the linear case, where $\boldsymbol{\varepsilon}$ is called the *model error* and $\boldsymbol{\eta}$ the *measurement error*. Both are random variables[6] which are often, although not necessarily, taken as zero-mean Gaussian. As a result of the random error terms, $\mathbf{x}$ and $\mathbf{y}$ also become random variables such that to completely quantify them it will be necessary to specify their probability distributions as functions of time. In the special case of linear equations, Gaussian error terms will result in Gaussian states and outputs which can be completely specified by their first and second moments (mean values and covariance matrices). In the more general, nonlinear, case it will be necessary to specify higher moments or ensembles of representative *realizations* of $\mathbf{x}$ and $\mathbf{y}$. In reservoir simulation it is not customary to introduce error terms that are additive to the states (as in Eq. (2.16)). Instead it is much more common to consider the parameters of the system equations, in particular the grid-block permeabilities, as uncertain. Typically these parameter uncertainties are considered to be so large that they dominate the model errors. Measurement errors are normally introduced in computer-assisted history matching; see e.g. Oliver et al. (2008). In this text we will not make use of error terms in the system description.

2.1.4 Linearized Equations

2.1.4.1 Jacobians

To analyze the nature of nonlinear system equations $\dot{\mathbf{x}} = \mathbf{f}(\mathbf{u}, \mathbf{x})$, or to approximate their solution through numerical computation, it is usually necessary to linearize them around a point in state-input space. Using a Taylor expansion and starting from Eq. (2.6) we can write:

$$\dot{\mathbf{x}} = \mathbf{f}(\mathbf{u}, \mathbf{x}) \approx \mathbf{f}(\mathbf{u}^0, \mathbf{x}^0) + \frac{\partial \mathbf{f}(\mathbf{u}^0, \mathbf{x}^0)}{\partial \mathbf{u}}(\mathbf{u} - \mathbf{u}^0) + \frac{\partial \mathbf{f}(\mathbf{u}^0, \mathbf{x}^0)}{\partial \mathbf{x}}(\mathbf{x} - \mathbf{x}^0)\,, \tag{2.18}$$

where we have neglected terms of second order and higher, and applied the usual short-cut notation

---

[6] The random model errors are also referred to as *random input*, or as a *stochastic forcing term*.

$$\frac{\partial \mathbf{f}(\mathbf{u}^0, \mathbf{x}^0)}{\partial \mathbf{x}} \triangleq \left. \frac{\partial \mathbf{f}(\mathbf{u}, \mathbf{x})}{\partial \mathbf{x}} \right|_{\mathbf{u}=\mathbf{u}^0, \mathbf{x}=\mathbf{x}^0} . \tag{2.19}$$

Defining

$$\bar{\mathbf{u}} \triangleq \mathbf{u} - \mathbf{u}^0, \tag{2.20}$$

$$\bar{\mathbf{x}} \triangleq \mathbf{x} - \mathbf{x}^0, \tag{2.21}$$

we can rewrite Eq. (2.18) as

$$\dot{\bar{\mathbf{x}}} + \dot{\mathbf{x}}^0 \approx \mathbf{f}(\mathbf{u}^0, \mathbf{x}^0) + \frac{\partial \mathbf{f}(\mathbf{u}^0, \mathbf{x}^0)}{\partial \mathbf{u}} \bar{\mathbf{u}} + \frac{\partial \mathbf{f}(\mathbf{u}^0, \mathbf{x}^0)}{\partial \mathbf{x}} \bar{\mathbf{x}}, \tag{2.22}$$

which, because

$$\dot{\mathbf{x}}^0 = \mathbf{f}(\mathbf{u}^0, \mathbf{x}^0), \tag{2.23}$$

can be reduced to the *linearized system equations*

$$\dot{\bar{\mathbf{x}}} = \bar{\mathbf{A}}(\mathbf{u}^0, \mathbf{x}^0) \bar{\mathbf{x}} + \bar{\mathbf{B}}(\mathbf{u}^0, \mathbf{x}^0) \bar{\mathbf{u}}, \tag{2.24}$$

where the *Jacobian matrices*[7] $\bar{\mathbf{A}}$ and $\bar{\mathbf{B}}$ are defined as

$$\bar{\mathbf{A}}(\mathbf{u}^0, \mathbf{x}^0) \triangleq \frac{\partial \mathbf{f}(\mathbf{u}^0, \mathbf{x}^0)}{\partial \mathbf{x}}, \tag{2.25}$$

$$\bar{\mathbf{B}}(\mathbf{u}^0, \mathbf{x}^0) \triangleq \frac{\partial \mathbf{f}(\mathbf{u}^0, \mathbf{x}^0)}{\partial \mathbf{u}}. \tag{2.26}$$

In a similar fashion we can linearize a nonlinear output function $\mathbf{y} = \mathbf{h}(\mathbf{u}, \mathbf{x})$ to obtain

$$\bar{\mathbf{y}} = \bar{\mathbf{C}}(\mathbf{u}^0, \mathbf{x}^0) \bar{\mathbf{x}} + \bar{\mathbf{D}}(\mathbf{u}^0, \mathbf{x}^0) \bar{\mathbf{u}}, \tag{2.27}$$

where the Jacobians $\bar{\mathbf{C}}$ and $\bar{\mathbf{D}}$ are defined as

$$\bar{\mathbf{C}}(\mathbf{u}^0, \mathbf{x}^0) \triangleq \frac{\partial \mathbf{h}(\mathbf{u}^0, \mathbf{x}^0)}{\partial \mathbf{x}}, \tag{2.28}$$

$$\bar{\mathbf{D}}(\mathbf{u}^0, \mathbf{x}^0) \triangleq \frac{\partial \mathbf{h}(\mathbf{u}^0, \mathbf{x}^0)}{\partial \mathbf{u}}. \tag{2.29}$$

If the system and output equations are given in implicit form $\mathbf{g}(\mathbf{u}, \mathbf{x}, \dot{\mathbf{x}}) = \mathbf{0}$ and $\mathbf{j}(\mathbf{u}, \mathbf{x}, \mathbf{y}) = \mathbf{0}$ we obtain linearized equations in terms of Jacobians

$$\bar{\bar{\mathbf{A}}} \triangleq \frac{\partial \mathbf{g}(\mathbf{u}, \mathbf{x}, \dot{\mathbf{x}})}{\partial \mathbf{x}}, \tag{2.30}$$

---

[7] Usually simply referred to as *Jacobians*.

$$\bar{\bar{\mathbf{B}}} \triangleq \frac{\partial \mathbf{g}(\mathbf{u}, \mathbf{x}, \dot{\mathbf{x}})}{\partial \mathbf{u}} , \tag{2.31}$$

$$\bar{\bar{\mathbf{E}}} \triangleq \frac{\partial \mathbf{g}(\mathbf{u}, \mathbf{x}, \dot{\mathbf{x}})}{\partial \dot{\mathbf{x}}} , \tag{2.32}$$

$$\bar{\bar{\mathbf{C}}} \triangleq \frac{\partial \mathbf{j}(\mathbf{u}, \mathbf{x}, \mathbf{y})}{\partial \mathbf{x}} \tag{2.33}$$

$$\bar{\bar{\mathbf{D}}} \triangleq \frac{\partial \mathbf{j}(\mathbf{u}, \mathbf{x}, \mathbf{y})}{\partial \mathbf{u}} , \tag{2.34}$$

$$\bar{\bar{\mathbf{F}}} \triangleq \frac{\partial \mathbf{g}(\mathbf{u}, \mathbf{x}, \mathbf{y})}{\partial \mathbf{y}} , \tag{2.35}$$

where we have dropped the superscripts 0 for clarity.

2.1.4.2 Secant and Tangent Matrices

In reservoir simulation one often encounters systems $\dot{\mathbf{x}} = \mathbf{f}(\mathbf{u}, \mathbf{x})$ that can be expressed in the form[8]

$$\dot{\mathbf{x}} = \mathbf{A}(\mathbf{x})\mathbf{x} + \mathbf{B}(\mathbf{x})\mathbf{u} . \tag{2.36}$$

In that case we obtain the linearized Eq. (2.24) with Jacobians defined as

$$\bar{\mathbf{A}}\left(\mathbf{u}^0, \mathbf{x}^0\right) \triangleq \mathbf{A}\left(\mathbf{x}^0\right) + \frac{\partial \mathbf{A}(\mathbf{x}^0)}{\partial \mathbf{x}} \mathbf{x}^0 + \frac{\partial \mathbf{B}(\mathbf{x}^0)}{\partial \mathbf{x}} \mathbf{u}^0 , \tag{2.37}$$

$$\bar{\mathbf{B}}\left(\mathbf{x}^0\right) \triangleq \mathbf{B}\left(\mathbf{x}^0\right) . \tag{2.38}$$

If we linearize the state equations along all points of a given trajectory $(\mathbf{x}^0(t), \mathbf{u}^0(t))$ in state-input space, the resulting model is referred to as the *tangent-linear* approximation of the nonlinear model, or simply the tangent-linear model. The Jacobians $\bar{\mathbf{A}}$ and $\bar{\mathbf{B}}$ are therefore also referred to as the *tangent matrices* of the system. Note that the matrices $\mathbf{A}$ and $\mathbf{B}$ are not tangent matrices because they do not describe the system dynamics tangent to the state trajectory. Instead they can be interpreted as *secant matrices*; see Fig. 2.1.

---

[8] In the systems-and-control literature this is known as a *control-affine* nonlinear equation. An affine function is a linear function plus a translation. Control affine functions are an important topic of study in nonlinear control theory.

Fig. 2.1 The secant and the tangent to a function $f(x)$ in a point $(x^0, f(x^0))$

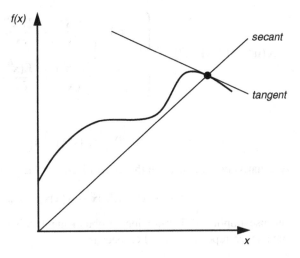

2.1.4.3 Generalized State-Space Form*

In reservoir simulation we often encounter systems that can be expressed in the generalized state-space form

$$\hat{\mathbf{E}}(\mathbf{x})\dot{\mathbf{x}} = \hat{\mathbf{A}}(\mathbf{x})\mathbf{x} + \hat{\mathbf{B}}(\mathbf{x})\mathbf{u}. \tag{2.39}$$

In that case we can linearize around a point $(\mathbf{u}^0, \mathbf{x}^0, \dot{\mathbf{x}}^0)$ to obtain the linearized equations

$$\bar{\hat{\mathbf{E}}}(\mathbf{x}^0)\dot{\bar{\mathbf{x}}} = \bar{\hat{\mathbf{A}}}(\mathbf{u}^0, \mathbf{x}^0, \dot{\mathbf{x}}^0)\bar{\mathbf{x}} + \bar{\hat{\mathbf{B}}}(\mathbf{x}^0)\bar{\mathbf{u}}, \tag{2.40}$$

with Jacobians defined as

$$\bar{\hat{\mathbf{A}}}(\mathbf{u}^0, \mathbf{x}^0, \dot{\mathbf{x}}^0) \triangleq \hat{\mathbf{A}}(\mathbf{x}^0) + \frac{\partial\hat{\mathbf{A}}(\mathbf{x}^0)}{\partial\mathbf{x}}\mathbf{x}^0 + \frac{\partial\hat{\mathbf{B}}(\mathbf{x}^0)}{\partial\mathbf{x}}\mathbf{u}^0 - \frac{\partial\hat{\mathbf{E}}(\mathbf{x}^0)}{\partial\mathbf{x}}\dot{\mathbf{x}}^0, \tag{2.41}$$

$$\bar{\hat{\mathbf{B}}}(\mathbf{x}^0) \triangleq \hat{\mathbf{B}}(\mathbf{x}^0), \tag{2.42}$$

$$\bar{\hat{\mathbf{E}}}(\mathbf{x}^0) \triangleq \hat{\mathbf{E}}(\mathbf{x}^0). \tag{2.43}$$

However, for analysis purposes it is normally more useful to bring this equation in the linearized ordinary state-space form (2.24):

$$\dot{\mathbf{x}} = \mathbf{f}(\mathbf{u}, \mathbf{x}) = \underbrace{\left(\hat{\mathbf{E}}(\mathbf{x})\right)^{-1}\hat{\mathbf{A}}(\mathbf{x})}_{\mathbf{A}(\mathbf{x})}\mathbf{x} + \underbrace{\left(\hat{\mathbf{E}}(\mathbf{x})\right)^{-1}\hat{\mathbf{B}}(\mathbf{x})}_{\mathbf{B}(\mathbf{x})}\mathbf{u}, \tag{2.44}$$

in which case the Jacobians $\bar{\mathbf{A}} = \partial\mathbf{f}(\mathbf{u}, \mathbf{x})/\partial\mathbf{x}$ and $\bar{\mathbf{B}} = \partial\mathbf{f}(\mathbf{u}, \mathbf{x})/\partial\mathbf{u}$ can be obtained as

$$\bar{\mathbf{A}}(\mathbf{u}^0,\mathbf{x}^0) \triangleq (\hat{\mathbf{E}}(\mathbf{x}^0))^{-1} \left\{ \begin{array}{l} \hat{\mathbf{A}}(\mathbf{x}^0) + \left[\dfrac{\partial \hat{\mathbf{A}}(\mathbf{x}^0)}{\partial \mathbf{x}} - \dfrac{\partial \hat{\mathbf{E}}(\mathbf{x}^0)}{\partial \mathbf{x}} (\hat{\mathbf{E}}(\mathbf{x}^0))^{-1} \hat{\mathbf{A}}(\mathbf{x}^0) \right] \mathbf{x}^0 \\[4mm] + \left[\dfrac{\partial \hat{\mathbf{B}}(\mathbf{x}^0)}{\partial \mathbf{x}} - \dfrac{\partial \hat{\mathbf{E}}(\mathbf{x}^0)}{\partial \mathbf{x}} (\hat{\mathbf{E}}(\mathbf{x}^0))^{-1} \hat{\mathbf{B}}(\mathbf{x}^0) \right] \mathbf{u}^0 \end{array} \right\},$$

$$\tag{2.45}$$

$$\bar{\mathbf{B}}(\mathbf{x}^0) \triangleq (\hat{\mathbf{E}}(\mathbf{x}^0))^{-1} \mathbf{B}(\mathbf{x}^0). \tag{2.46}$$

Alternatively we can write the generalized state Eq. (2.39) in implicit form

$$\hat{\mathbf{g}}(\mathbf{u},\mathbf{x},\dot{\mathbf{x}}) \triangleq \hat{\mathbf{E}}(\mathbf{x})\dot{\mathbf{x}} - \hat{\mathbf{A}}(\mathbf{x})\mathbf{x} - \hat{\mathbf{B}}(\mathbf{x})\mathbf{u} = \mathbf{0}, \tag{2.47}$$

and use implicit differentiation to obtain the Jacobian $\bar{\mathbf{A}}$ related to the ordinary state-space representation. I.e., because

$$\frac{d\hat{\mathbf{g}}}{d\mathbf{x}} = \frac{\partial \hat{\mathbf{g}}}{\partial \mathbf{x}} + \frac{\partial \hat{\mathbf{g}}}{\partial \dot{\mathbf{x}}} \frac{\partial \dot{\mathbf{x}}}{\partial \mathbf{x}} = \mathbf{0}, \tag{2.48}$$

we have

$$\frac{\partial \dot{\mathbf{x}}}{\partial \mathbf{x}} = -\left(\frac{\partial \hat{\mathbf{g}}}{\partial \dot{\mathbf{x}}} \right)^{-1} \frac{\partial \hat{\mathbf{g}}}{\partial \mathbf{x}}, \tag{2.49}$$

and because $\bar{\mathbf{A}} = \partial \mathbf{f}(\mathbf{u},\mathbf{x})/\partial \mathbf{x} = \partial \dot{\mathbf{x}}/\partial \mathbf{x}$ we find that

$$\bar{\mathbf{A}}(\mathbf{u}^0,\mathbf{x}^0,\dot{\mathbf{x}}^0) = (\hat{\mathbf{E}}(\mathbf{x}^0))^{-1} \left[\hat{\mathbf{A}}(\mathbf{x}^0) + \frac{\partial \hat{\mathbf{A}}(\mathbf{x}^0)}{\partial \mathbf{x}} \mathbf{x}^0 + \frac{\partial \hat{\mathbf{B}}(\mathbf{x}^0)}{\partial \mathbf{x}} \mathbf{u}^0 - \frac{\partial \hat{\mathbf{E}}(\mathbf{x}^0)}{\partial \mathbf{x}} \dot{\mathbf{x}}^0 \right],$$

$$\tag{2.50}$$

which, with the aid of Eqs. (2.41) and (2.43), can also be expressed as

$$\bar{\mathbf{A}}(\mathbf{u}^0,\mathbf{x}^0,\dot{\mathbf{x}}^0) = \left(\bar{\hat{\mathbf{E}}}(\mathbf{x}^0) \right)^{-1} \bar{\hat{\mathbf{A}}}(\mathbf{u}^0,\mathbf{x}^0,\dot{\mathbf{x}}^0). \tag{2.51}$$

2.2   Single-Phase Flow

2.2.1   System Equations

As a first application, we consider flow of a weakly-compressible single-phase liquid through a porous medium. The derivation of the governing PDEs and the semi-discretization has been presented in Chap. 1. We used a finite difference discretization, but the following theory is equally applicable to results derived with other semi-discretization methods. Use of any of the methods produces a system of ODEs that can be written in matrix form as:

$$\mathbf{V}\dot{\mathbf{p}} + \mathbf{T}\mathbf{p} = \mathbf{q}. \tag{2.52}$$

Here $\mathbf{V}$ and $\mathbf{T}$ are matrices with entries that depend on dynamic and static reservoir properties, $\mathbf{p}$ is a vector of pressures and $\mathbf{q}$ is a vector of volumetric flow rates. $\mathbf{V}$ is a diagonal matrix known as the *accumulation matrix* and $\mathbf{T}$ is a symmetric banded matrix, known as the *transmissibility matrix*. The flow rates $\mathbf{q}$ correspond to flow into or out of the reservoir, i.e. to wells, and are expressed in m^3/s. Positive values imply injection and negative values imply production. Because usually only a few grid blocks are penetrated by wells, only a few elements of $\mathbf{q}$ have a non-zero value. In the case of a reservoir modeled with n grid blocks and produced with m wells, $\mathbf{V}$ and $\mathbf{T}$ would be $n \times n$ matrices, and $\mathbf{p}$ and $\mathbf{q}$ would be $n \times 1$ vectors, of which $\mathbf{q}$ would have m non-zero entries. Equation (2.52) can be re-casted in state variable form (2.8) through definition of

$$\mathbf{u} \triangleq \mathbf{L}_{uq}\mathbf{q}, \tag{2.53}$$

$$\mathbf{x} \triangleq \mathbf{p}. \tag{2.54}$$

In single-phase flow the state variables $\mathbf{x}$ are just identical to the pressures $\mathbf{p}$. The vector $\mathbf{u}$ represents the inputs to the system, which are in our case the non-zero elements of the flow rate vector $\mathbf{q}$. The matrix $\mathbf{L}_{uq}$ is therefore a *location matrix*, also known as a *selection matrix* which contains only ones and zeros at the appropriate places. The inverse relationship is given by

$$\mathbf{q} = \mathbf{L}_{qu}\mathbf{u}, \tag{2.55}$$

where

$$\mathbf{L}_{qu} = \mathbf{L}_{uq}^T, \tag{2.56}$$

Substitution of relationships (2.53) and (2.55) in Eq. (2.52) results in the generalized state-space representation (2.9) with matrices $\hat{\mathbf{A}}$, $\hat{\mathbf{B}}$ and $\hat{\mathbf{E}}$ given by:

$$\hat{\mathbf{A}} \triangleq -\mathbf{T}, \tag{2.57}$$

$$\hat{\mathbf{B}} \triangleq \mathbf{L}_{qu}, \tag{2.58}$$

$$\hat{\mathbf{E}} \triangleq \mathbf{V}. \tag{2.59}$$

The ordinary state-space form (2.8) is obtained by defining the matrices $\mathbf{A}$ and $\mathbf{B}$ as[9]

---

[9] In a numerical implementation, the inverse $\mathbf{V}^{-1}$ of the diagonal matrix $\mathbf{V}$ can be computed very efficiently by just taking the reciprocals of the diagonal elements.

$$\mathbf{A} \triangleq -\mathbf{V}^{-1}\mathbf{T}, \tag{2.60}$$

$$\mathbf{B} \triangleq \mathbf{V}^{-1}\mathbf{L}_{qu}. \tag{2.61}$$

If we choose the output vector **y** to consist of only those pressures that are accessible to measurements, the matrix **C** is therefore also a selection matrix. Matrix **D** is zero because there is no direct dependency of the output on the input. In reality the outputs are usually surface measurements of the tubing-head pressure, and therefore we should include a description of the dynamic behavior of the well between the reservoir and the surface. However, as a first assumption, we neglect well dynamics and assume that the wells are equipped with permanent down hole gauges (PDGs) to measure the pressures. In the case of a reservoir modeled with n grid blocks and produced with m wells, of which m_p contain PDGs, matrices **A** and **B** have dimension $n \times n$, matrix **C** dimension $m_p \times n$, and vectors **u**, **x** and **y** dimensions $m \times 1$, $n \times 1$ and $m_p \times 1$ respectively. Matrix Eq. (2.8) represents a system of linear first order ODEs with constant coefficients, i.e. an LTI system. Starting from an initial value $\breve{\mathbf{x}}$, the ODEs for **x** can be integrated in time, and because the equations are linear the solution can be expressed analytically. Alternatively, the integration can be performed numerically as will be discussed in Chap. 3.

2.2.2 Example 1 Continued—Location Matrix

Reconsidering the six-grid-block example introduced in Sect. 1.3.3, the location matrix $\mathbf{L}_{uq}$ as defined in Eq. (2.53) is given by

$$\underbrace{\begin{bmatrix} u_1 \\ u_2 \end{bmatrix}}_{\mathbf{u}} = \underbrace{\begin{bmatrix} 1 & 0 & 0 & 0 & 0 & 0 \\ 0 & 0 & 0 & 0 & 0 & 1 \end{bmatrix}}_{\mathbf{L}_{uq}} \underbrace{\begin{bmatrix} q_1 \\ 0 \\ 0 \\ 0 \\ 0 \\ q_6 \end{bmatrix}}_{\mathbf{q}}. \tag{2.62}$$

If the output **y** consists of the pressures in the two wells, the output matrix $\mathbf{C} = \mathbf{L}_{qu} = \mathbf{L}_{uq}^T$ is given by:

$$\underbrace{\begin{bmatrix} y_1 \\ y_2 \end{bmatrix}}_{\mathbf{y}} = \underbrace{\begin{bmatrix} 1 & 0 & 0 & 0 & 0 & 0 \\ 0 & 0 & 0 & 0 & 0 & 1 \end{bmatrix}}_{\mathbf{C}} \underbrace{\begin{bmatrix} x_1 \\ x_2 \\ x_3 \\ x_4 \\ x_5 \\ x_6 \end{bmatrix}}_{\mathbf{x}}. \tag{2.63}$$

2.2.3 Prescribed Pressures and Flow Rates

Until now we have assumed that the source terms, i.e. the flow rates in the wells, were the input variables, and that their values can be prescribed as a function of time. However, it is also possible to control the system by prescribing the state variables, i.e. the pressures in the wells. Note that it is not possible to prescribe both pressure and flow rate in a well; either one of them should be fixed and the other left free, or a mathematical relationship between them should be specified which may be in algebraic or differential form. The most commonly used method in reservoir engineering is through the definition of a *well model*, which is an algebraic relationship between the grid-block pressure and the well flow rate. Alternatively, one of the pressures may be prescribed directly, resulting in a reduction of the length of the state vector with one element. We will discuss both methods in the following sections. In order to take into account prescribed pressures and flow rates in a structured way it is convenient to re-order the variables in Eq. (2.52) such that the prescribed and the non-prescribed, free, values are grouped. In addition, we take the opportunity to make a distinction between prescribed flow rates in grid blocks with and without wells.[10] We can formally describe the re-ordering with the aid of a *permutation matrix* $\mathbf{L}$ as:

$$\mathbf{p}^* \triangleq \begin{bmatrix} \mathbf{p}_1^* \\ \mathbf{p}_2^* \\ \mathbf{p}_3^* \end{bmatrix} \equiv \mathbf{L}_{p*p}\mathbf{p}\,, \tag{2.64}$$

where $\mathbf{p}_1^*$ are the pressures in the grid blocks that are not penetrated by a well, i.e. where the source terms $\mathbf{q}_1^*$ are equal to zero, $\mathbf{p}_2^*$ are the pressures in the blocks where the source terms $\mathbf{q}_2^*$ are prescribed as well flow rates, and $\mathbf{p}_3^*$ are the pressures in the blocks where the source terms $\mathbf{q}_3^*$ are obtained through prescription of the bottom-hole pressures in the wells. Similarly we can write

$$\mathbf{q}^* \triangleq \begin{bmatrix} \mathbf{0} \\ \mathbf{q}_2^* \\ \mathbf{q}_3^* \end{bmatrix} \equiv \mathbf{L}_{q*q}\mathbf{q}\,, \tag{2.65}$$

where we choose

$$\mathbf{L}_{q*q} = \mathbf{L}_{p*p}\,, \tag{2.66}$$

which means that we re-order the elements of $\mathbf{p}$ and $\mathbf{q}$ in Eq. (2.52) identically, i.e. we interchange the rows of the equations. The permutation matrix $\mathbf{L}_{p*p} = \mathbf{L}_{q*q}$ is an identity matrix with interchanged rows. Permutation matrices are orthogonal, which implies that

---

[10] In grid blocks that are not penetrated by a well the prescribed flow rates are of course equal to zero.

$$\mathbf{L}_{p*p}\mathbf{L}^T_{p*p} = \mathbf{I}. \qquad (2.67)$$

The inverse relationships corresponding to expressions (2.64) and (2.65) are therefore given by

$$\mathbf{p} = \mathbf{L}_{pp*}\mathbf{p}^*, \qquad (2.68)$$

$$\mathbf{q} = \mathbf{L}_{qq*}\mathbf{q}^*, \qquad (2.69)$$

where

$$\mathbf{L}_{pp*} = \mathbf{L}_{qq*} = \mathbf{L}^T_{p*p} = \mathbf{L}^T_{q*q}. \qquad (2.70)$$

Substitution of Eqs. (2.68) and (2.69) in Eq. (2.52) and reorganizing the terms results in

$$\mathbf{V}^*\dot{\mathbf{p}}^* + \mathbf{T}^*\mathbf{p}^* = \mathbf{q}^*, \qquad (2.71)$$

where $\mathbf{T}^*$ and $\mathbf{V}^*$ are given by

$$\mathbf{T}^* = \mathbf{L}_{q*q}\mathbf{T}\mathbf{L}_{pp*}, \qquad (2.72)$$

$$\mathbf{V}^* = \mathbf{L}_{q*q}\mathbf{V}\mathbf{L}_{pp*}. \qquad (2.73)$$

Equation (2.71) can be written in partitioned form as:

$$\begin{bmatrix} \mathbf{V}^*_{11} & \mathbf{0} & \mathbf{0} \\ \mathbf{0} & \mathbf{V}^*_{22} & \mathbf{0} \\ \mathbf{0} & \mathbf{0} & \mathbf{V}^*_{33} \end{bmatrix} \begin{bmatrix} \dot{\mathbf{p}}^*_1 \\ \dot{\mathbf{p}}^*_2 \\ \dot{\mathbf{p}}^*_3 \end{bmatrix} + \begin{bmatrix} \mathbf{T}^*_{11} & \mathbf{T}^*_{12} & \mathbf{T}^*_{13} \\ \mathbf{T}^*_{21} & \mathbf{T}^*_{22} & \mathbf{T}^*_{23} \\ \mathbf{T}^*_{31} & \mathbf{T}^*_{32} & \mathbf{T}^*_{33} \end{bmatrix} \begin{bmatrix} \mathbf{p}^*_1 \\ \mathbf{p}^*_2 \\ \mathbf{p}^*_3 \end{bmatrix} = \begin{bmatrix} \mathbf{0} \\ \mathbf{q}^*_2 \\ \mathbf{q}^*_3 \end{bmatrix}. \qquad (2.74)$$

Note that the diagonal structure of matrix $\mathbf{V}$ has been maintained in $\mathbf{V}^*$. We can also apply the partitioning to the state-space representation, in which case we may choose not to partition $\mathbf{u}$ and $\mathbf{y}$, or to partition them also. We choose to partition them, according to

$$\mathbf{u}^* \triangleq \mathbf{L}_{u*u}\mathbf{u}, \qquad (2.75)$$

$$\mathbf{y}^* \triangleq \mathbf{L}_{y*y}\mathbf{y}, \qquad (2.76)$$

where details of the partitioning are left open for the moment. Substitution of Eqs. (2.68), (2.69) and the inverse of Eq. (2.75) in Eqs. (2.53) and (2.54) results in

$$\mathbf{u}^* = \mathbf{L}_{u*q*}\mathbf{q}^*, \qquad (2.77)$$

$$\mathbf{x}^* = \mathbf{p}^*, \qquad (2.78)$$

where

$$\mathbf{L}_{u*q*} = \mathbf{L}_{u*u}\mathbf{L}_{uq}\mathbf{L}_{qq*}, \qquad (2.79)$$

$$\mathbf{x}^* = \mathbf{L}_{p*p}\mathbf{x}. \qquad (2.80)$$

The partitioned state-space representation can then be written as

$$\dot{\mathbf{x}}^* = \mathbf{A}^*\mathbf{x}^* + \mathbf{B}^*\mathbf{u}^* , \qquad (2.81)$$

$$\mathbf{y}^* = \mathbf{C}^*\mathbf{x}^* , \qquad (2.82)$$

where

$$\mathbf{A}^* \triangleq - (\mathbf{V}^*)^{-1}\mathbf{T}^* = \mathbf{L}_{q*q}\mathbf{A}\mathbf{L}_{pp*} , \qquad (2.83)$$

$$\mathbf{B}^* \triangleq (\mathbf{V}^*)^{-1}\mathbf{L}_{q^*u^*} = \mathbf{L}_{q*q}\mathbf{B}\mathbf{L}_{uu*} , \qquad (2.84)$$

$$\mathbf{C}^* \triangleq \mathbf{L}_{y*y}\mathbf{C}\mathbf{L}_{pp*} . \qquad (2.85)$$

The re-ordering of vector and matrix elements using permutation matrices as described above is a formal technique. It results in partitioned vectors and matrices that allow for a structured handling of prescribed pressures. However, for a numerical implementation it is not essential to actually perform the re-ordering. In the following we will therefore omit the star superscripts and simply work with partitioned matrices without the use of permutation matrices.

2.2.4 Well Models

2.2.4.1 Prescribed Bottom-Hole Pressures and Well Flow Rates

The standard approach in reservoir simulation to prescribe bottom-hole pressures is through the definition of a *well model*. In that case the flow rate q in the grid block where we want to prescribe the pressure is defined as

$$q = J_{well}\left(\breve{p}_{well} - p\right) , \qquad (2.86)$$

where $\breve{p}_{well}$ is the prescribed bottom-hole pressure, p is the grid-block pressure and J_{well} is called the well index or productivity index. The well index is a function of the grid-block geometry and reflects the effect of near-well flow which is normally not properly represented by the finite-difference discretization because the grid block dimensions are usually much larger than the well diameter; see also Sect. 1.3.6. Note that, in line with our convention, a negative flow rate indicates production. Use of Eq. (2.86) can be interpreted as specifying an algebraic relationship between the state variable (i.e. the pressure) and the source term (i.e. the flow rate) in the grid block that contains the well. Equation (2.86) can be generalized for multiple wells to

$$\mathbf{q}_3 = \mathbf{J}_3\left(\breve{p}_{well} - \mathbf{p}_3\right) , \qquad (2.87)$$

where $\mathbf{J}_3$ is a diagonal matrix of well indices J_{well}, and $\breve{\mathbf{p}}_{well}$ is a vector of pre-scribed bottom-hole pressures. In a similar fashion, we can write

$$\mathbf{q}_2 = \breve{\mathbf{q}}_{well}, \tag{2.88}$$

where $\breve{\mathbf{q}}_{well}$ are the prescribed well rates. If we combine Eqs. (2.87) and (2.88) with Eq. (2.74), and reorganize terms, we obtain

$$
\begin{bmatrix} \mathbf{V}_{11} & \mathbf{0} & \mathbf{0} \\ \mathbf{0} & \mathbf{V}_{22} & \mathbf{0} \\ \mathbf{0} & \mathbf{0} & \mathbf{V}_{33} \end{bmatrix} \begin{bmatrix} \dot{\mathbf{p}}_1 \\ \dot{\mathbf{p}}_2 \\ \dot{\mathbf{p}}_3 \end{bmatrix} + \begin{bmatrix} \mathbf{T}_{11} & \mathbf{T}_{12} & \mathbf{T}_{13} \\ \mathbf{T}_{21} & \mathbf{T}_{22} & \mathbf{T}_{23} \\ \mathbf{T}_{31} & \mathbf{T}_{32} & \mathbf{T}_{33} + \mathbf{J}_3 \end{bmatrix} \begin{bmatrix} \mathbf{p}_1 \\ \mathbf{p}_2 \\ \mathbf{p}_3 \end{bmatrix} = \begin{bmatrix} \mathbf{0} \\ \breve{\mathbf{q}}_{well} \\ \mathbf{J}_3 \breve{\mathbf{p}}_{well} \end{bmatrix}.
$$
$$\tag{2.89}$$

An important aspect of the introduction of the well model is that, compared to matrix $\mathbf{T}$ in Eq. (2.52), the transmissibility matrix in Eq. (2.89) has elements added to its main diagonal. We will discuss the consequences of this addition in Chap. 3.

2.2.4.2   Free Bottom-Hole Pressures and Well Flow Rates

The flow rates $\bar{\mathbf{q}}_{well} = \mathbf{q}_3$ in the wells where the bottom-hole pressures have been prescribed can be obtained directly from Eq. (2.87) as

$$\bar{\mathbf{q}}_{well} = \mathbf{J}_3 \left(\breve{\mathbf{p}}_{well} - \mathbf{p}_3 \right). \tag{2.90}$$

To compute the bottom-hole pressures $\bar{\mathbf{p}}_{well}$ in the wells where the flow rates have been prescribed we need an additional diagonal matrix $\mathbf{J}_{q,2}$ of well indices J_q. We can then write

$$\breve{\mathbf{q}}_{well} = \mathbf{J}_2 (\bar{\mathbf{p}}_{well} - \mathbf{p}_2), \tag{2.91}$$

from which we obtain

$$\bar{\mathbf{p}}_{well} = \mathbf{p}_2 + \mathbf{J}_2^{-1} \breve{\mathbf{q}}_{well}. \tag{2.92}$$

2.2.4.3   State-Space Representation

If we define the (partitioned) state, input and output vectors

$$\mathbf{x} \triangleq \mathbf{p} = \begin{bmatrix} \mathbf{p}_1 \\ \mathbf{p}_2 \\ \mathbf{p}_3 \end{bmatrix}, \tag{2.93}$$

$$\mathbf{u} \triangleq \begin{bmatrix} \breve{\mathbf{q}}_{well} \\ \breve{\mathbf{p}}_{well} \end{bmatrix}, \tag{2.94}$$

$$\mathbf{y} \triangleq \begin{bmatrix} \bar{\mathbf{p}}_{well} \\ \bar{\mathbf{q}}_{well} \end{bmatrix}, \tag{2.95}$$

equations (2.89), (2.90) and (2.92) can be rewritten in state-space form as

$$\dot{\mathbf{x}} = \mathbf{A}\mathbf{x} + \mathbf{B}\mathbf{u}, \tag{2.96}$$

$$\mathbf{y} = \mathbf{C}\mathbf{x} + \mathbf{D}\mathbf{u}, \tag{2.97}$$

where the matrices are defined as

$$\mathbf{A} \triangleq - \begin{bmatrix} \mathbf{V}_{11}^{-1}\mathbf{T}_{11} & \mathbf{V}_{11}^{-1}\mathbf{T}_{12} & \mathbf{V}_{11}^{-1}\mathbf{T}_{13} \\ \mathbf{V}_{22}^{-1}\mathbf{T}_{21} & \mathbf{V}_{22}^{-1}\mathbf{T}_{22} & \mathbf{V}_{22}^{-1}\mathbf{T}_{23} \\ \mathbf{V}_{33}^{-1}\mathbf{T}_{31} & \mathbf{V}_{33}^{-1}\mathbf{T}_{32} & \mathbf{V}_{33}^{-1}(\mathbf{T}_{33}+\mathbf{J}_3) \end{bmatrix}, \tag{2.98}$$

$$\mathbf{B} \triangleq \begin{bmatrix} \mathbf{0} & \mathbf{0} \\ \mathbf{V}_{22}^{-1} & \mathbf{0} \\ \mathbf{0} & \mathbf{V}_{33}^{-1}\mathbf{J}_3 \end{bmatrix}, \tag{2.99}$$

$$\mathbf{C} \triangleq \begin{bmatrix} \mathbf{0} & \mathbf{I} & \mathbf{0} \\ \mathbf{0} & \mathbf{0} & -\mathbf{J}_3 \end{bmatrix}, \tag{2.100}$$

$$\mathbf{D} \triangleq \begin{bmatrix} \mathbf{J}_2^{-1} & \mathbf{0} \\ \mathbf{0} & \mathbf{J}_3 \end{bmatrix}. \tag{2.101}$$

2.2.5 Example 1 Continued—Well Model

In Example 1, discussed in Sec. 2.2.2, we fixed the flow rates in both wells. Here, we fix the bottom-hole pressure of the producer in grid block 6 as: $p_{well} = 28.00 \times 10^6$ Pa (4061 psi), while we choose an injection rate in block 1 as $q_1 = 0.01$ m³/s (864 m³/d, 5434 bpd), where we use the convention that positive flow rates indicate injection. Because we only have one well with a prescribed pressure and one with a prescribed rate, we have

$$\breve{\mathbf{p}}_{well} = \begin{bmatrix} 28.00 \times 10^6 \end{bmatrix}, \tag{2.102}$$

$$\breve{\mathbf{q}}_{well} = [0.01]. \tag{2.103}$$

Correspondingly, the matrices $\mathbf{J}_3$ and $\mathbf{J}_2$ contain only one element each. Using the data for the near-well permeabilities as derived in Sect. 1.3.6 they become

$$\mathbf{J}_3 = \begin{bmatrix} 3.72 \times 10^{-9} \end{bmatrix}, \tag{2.104}$$

$$\mathbf{J}_2 = \begin{bmatrix} 3.72 \times 10^{-8} \end{bmatrix}. \tag{2.105}$$

2.2.6 Elimination of Prescribed Pressures*

An alternative way to implement a prescribed pressure is through directly pre-scribing the grid-block pressure. This means that one of the state variables is fixed, and may be eliminated from the system equations. To illustrate this method, we start again from the partitioned system Eq. (2.74). We indicate prescribed values with a '˅' above the variable, and free values with a '–':

$$\begin{bmatrix} \mathbf{V}_{11} & \mathbf{0} & \mathbf{0} \\ \mathbf{0} & \mathbf{V}_{22} & \mathbf{0} \\ \mathbf{0} & \mathbf{0} & \mathbf{V}_{33} \end{bmatrix} \begin{bmatrix} \dot{\bar{\mathbf{p}}}_1 \\ \dot{\bar{\mathbf{p}}}_2 \\ \dot{\check{\mathbf{p}}}_3 \end{bmatrix} + \begin{bmatrix} \mathbf{T}_{11} & \mathbf{T}_{12} & \mathbf{T}_{13} \\ \mathbf{T}_{21} & \mathbf{T}_{22} & \mathbf{T}_{23} \\ \mathbf{T}_{31} & \mathbf{T}_{32} & \mathbf{T}_{33} \end{bmatrix} \begin{bmatrix} \bar{\mathbf{p}}_1 \\ \bar{\mathbf{p}}_2 \\ \check{\mathbf{p}}_3 \end{bmatrix} = \begin{bmatrix} \mathbf{0} \\ \check{\mathbf{q}}_2 \\ \bar{\mathbf{q}}_3 \end{bmatrix}. \tag{2.106}$$

From the first two rows of matrix Eq. (2.106) we find the system of ODEs

$$\begin{bmatrix} \mathbf{V}_{11} & \mathbf{0} \\ \mathbf{0} & \mathbf{V}_{22} \end{bmatrix} \begin{bmatrix} \dot{\bar{\mathbf{p}}}_1 \\ \dot{\bar{\mathbf{p}}}_2 \end{bmatrix} = - \begin{bmatrix} \mathbf{T}_{11} & \mathbf{T}_{12} \\ \mathbf{T}_{21} & \mathbf{T}_{22} \end{bmatrix} \begin{bmatrix} \bar{\mathbf{p}}_1 \\ \bar{\mathbf{p}}_2 \end{bmatrix} \underbrace{- \begin{bmatrix} \mathbf{T}_{13} \\ \mathbf{T}_{23} \end{bmatrix} \check{\mathbf{p}}_3 + \begin{bmatrix} \mathbf{0} \\ \check{\mathbf{q}}_2 \end{bmatrix}}_{\text{prescribed}}. \tag{2.107}$$

Because we eliminated the prescribed pressures, the length of the pressure vector has been reduced. From the third row of Eq. (2.106) it follows that

$$\bar{\mathbf{q}}_3 = \begin{bmatrix} \mathbf{T}_{31} & \mathbf{T}_{32} \end{bmatrix} \begin{bmatrix} \bar{\mathbf{p}}_1 \\ \bar{\mathbf{p}}_2 \end{bmatrix} + \underbrace{\mathbf{T}_{33} \check{\mathbf{p}}_3 + \mathbf{V}_{33} \dot{\check{\mathbf{p}}}}_{\text{prescribed}}, \tag{2.108}$$

where $\bar{\mathbf{q}}_3$ represents the free flow rates in the wells where the pressures have been prescribed. Apparently the price to pay for the reduced length of the pressure vector is an increase in the number of input parameters to compute the free flow rates in case of time-varying prescribed pressures. Equations (2.107) and (2.108) can be rewritten in partitioned state-space form, as in Eqs. (2.96) and (2.97), through definition of

$$\mathbf{A} \triangleq - \begin{bmatrix} \mathbf{V}_{11}^{-1} & \mathbf{0} \\ \mathbf{0} & \mathbf{V}_{22}^{-1} \end{bmatrix} \begin{bmatrix} \mathbf{T}_{11} & \mathbf{T}_{12} \\ \mathbf{T}_{21} & \mathbf{T}_{22} \end{bmatrix}, \tag{2.109}$$

$$\mathbf{x} \triangleq \begin{bmatrix} \dot{\bar{\mathbf{p}}}_1 \\ \dot{\bar{\mathbf{p}}}_2 \end{bmatrix}, \tag{2.110}$$

$$\mathbf{B} \triangleq \begin{bmatrix} \mathbf{0} & -\mathbf{V}_{11}^{-1}\mathbf{T}_{13} & \mathbf{0} \\ \mathbf{V}_{22}^{-1} & -\mathbf{V}_{22}^{-1}\mathbf{T}_{23} & \mathbf{0} \end{bmatrix}, \tag{2.111}$$

$$\mathbf{u} \triangleq \begin{bmatrix} \breve{\mathbf{q}}_2 \\ \breve{\mathbf{p}}_3 \\ \mathbf{p}_3 \end{bmatrix},$$ (2.112)

$$\mathbf{y} \triangleq \begin{bmatrix} \bar{\mathbf{p}}_2 \\ \bar{\mathbf{q}}_3 \end{bmatrix},$$ (2.113)

$$\mathbf{C} \triangleq \begin{bmatrix} \mathbf{0} & \mathbf{I} \\ \mathbf{T}_{31} & \mathbf{T}_{32} \end{bmatrix},$$ (2.114)

$$\mathbf{D} \triangleq \begin{bmatrix} \mathbf{0} & \mathbf{0} & \mathbf{0} \\ \mathbf{0} & \mathbf{T}_{33} & \mathbf{V}_{33} \end{bmatrix}.$$ (2.115)

As before, we have chosen the output vector $\mathbf{y}$ such that it contains the free pressures and flow rates in the wells. However the input vector $\mathbf{u}$ now not only contains the prescribed flow rates and pressures in the wells, but also the time derivatives of the pressures. This technique to eliminate the prescribed state variables is mainly of theoretical value, and is not commonly used, if at all, in reservoir engineering applications.

2.2.7 System Energy*

The *energy balance* in flow through porous media is governed by three components:

1) *Potential energy* in the form of compressed fluids inside compressed rock and in the form of elevated fluid mass.
2) *Energy dissipation* caused by resistance to fluids flowing through the pore network.
3) *Energy transport* through the system boundaries in the form of *work* done by injecting or producing fluids under a pressure differential in the wells.

Note that we do not consider *kinetic energy*, because of the assumption that inertia forces may be neglected due to the very low flow velocities inside the pores.[11] Moreover, we will not take into account the effect of elevation on the potential energy because we restrict the theory and examples to two-dimensional horizontal reservoirs where gravity forces can be neglected.[12] Finally, we maintain our earlier

---

[11] I.e. we do not consider kinetic energy at a macroscopic level. We do take into account energy dissipation, which is the change of mechanical energy into thermal energy, or heat, and which at an atomic level can be interpreted as kinetic energy again.

[12] Elevation-related potential energy plays an important role in well-bore flow . Most reservoirs have enough potential energy, at least initially, to *lift* the oil to surface naturally in the production wells. This lift effect is in most cases caused by the difference in density between oil and water,

assumption of isothermal conditions, which implies that the heat generated by energy dissipation is instantaneously transferred to the surroundings (i.e. to outside the reservoir boundaries) such that there is no increase in reservoir temperature.

2.2.7.1 Potential Energy*

Starting from the assumption of a small, pressure-independent, total compressibility c_t, the fluid volume $\bar{V}(p)$ in a single grid block with volume V is expressed as[13]

$$\bar{V}(p) = V\phi_0[1 + c_t(p - p_0)] , \qquad (2.116)$$

where p_0 is a reference pressure, and ϕ_0 the corresponding porosity. The difference in potential energy when the fluids in the grid block experience a pressure increase from p_0 to p can therefore be expressed as

$$E_{pot}(p) = \int_{p_0}^{p} \frac{\partial \bar{V}(\pi)}{\partial \pi} \pi d\pi = \int_{p_0}^{p} V\phi_0 c_t \pi d\pi = \frac{1}{2} V\phi_0 c_t \left(p^2 - p_0^2\right) . \qquad (2.117)$$

If we choose the reference value for E_{pot} as zero at the reference pressure (which we may conveniently take to be the initial reservoir pressure p_R), we can compute the total potential energy in a reservoir model through summation over the grid blocks according to

$$E_{pot,tot}(t) = \sum_{i=1}^{n_{gb}} \frac{1}{2} V_i \phi_i c_t p_i^2(t) , \qquad (2.118)$$

where n_{gb} is the total number of grid blocks, and where V_i, ϕ_i and p_i are the grid-block volumes, porosities and pressures respectively, with only the pressures being a function of time.

2.2.7.2 Dissipation Energy*

The energy dissipated per unit time by resistance to flow through a grid-block boundary can be expressed as

$$\frac{dE_{dis}}{dt} = -\tilde{q}\Delta p = T\Delta p^2 , \qquad (2.119)$$

(Footnote 12 continued)
such that in an oil-filled well that drains a hydrostatically-pressured reservoir the oil will be lifted to surface because of elevation-related potential energy.

[13] For a detailed derivation of pressure-related potential energy for the case of pressure-dependent rock and fluid compressibilities, see Hubbert (1940).

where $\tilde{q}$ is the volumetric flow rate, Δp is the pressure difference between the two grid-block centers, and T is the grid-block transmissibility as defined in Eq. (1.30). In addition to dissipation at the grid-block boundaries, a large amount of energy is dissipated in the near-well-bore region where large pressure gradients are present. The energy dissipated per unit time by resistance to flow in a well grid block can be expressed as

$$\frac{dE_{dis}}{dt} = q_{well}\left(p_{well} - p_{gb}\right) = J_{well}\left(p_{well} - p_{gb}\right)^2, \qquad (2.120)$$

where q_{well} is the well flow rate (positive for injection), p_{well} is the flowing bottom-hole pressure, p_{gb} is the well-grid-block pressure and J_{well} is the well index, as defined in Eq. (1.44). The total amount of energy dissipated per unit time in a reservoir model is therefore obtained through summation over all grid-block connectivities and all wells as

$$\frac{dE_{dis,tot}(t)}{dt} = -\sum_{j=1}^{n_{con}} T_j(t)\Delta p_j^2(t) + \sum_{k=1}^{n_{well}} J_{well,k}(t)\left[p_{well,k}(t) - p_{gb,k}(t)\right]^2, \qquad (2.121)$$

where n_{con} is the number of connectivities, n_{well} is the number of wells, and where the transmissibilities, well indices and pressure drops may be functions of time.

2.2.7.3 Work Done by Wells*

The *power* (i.e. the work per unit time[14]) delivered by fluids injected into the reservoir can be expressed as

$$P_{well} = q_{well}p_{well}. \qquad (2.122)$$

The same equation holds for production wells, where we use the convention that flow rates in the producers have negative values. The total power delivered by all injectors and producers to the reservoir is therefore given by

$$P_{well,tot}(t) = \sum_{k=1}^{n_{well}} q_{well,k}(t) \times p_{well,k}(t). \qquad (2.123)$$

2.2.7.4 Energy Balance*

The total energy balance for the reservoir over a time interval $\Delta t = t_2 - t_1$ can now be expressed as

---

[14] Energy delivered to a system through mechanical or hydraulic action is often referred to as *work*. Work (or energy) per unit time is then called *power*. In strict SI units time is expressed in s (seconds), energy and work in J (Joule) and power in W (Watt), such that 1 W is equal to 1 J/s.

$$E_{pot,tot}(t_2) - E_{pot,tot}(t_1) + \underbrace{\int_{t_1}^{t_2} \frac{dE_{dis,tot}(t)}{dt}\, dt}_{dissipation} = \underbrace{\int_{t_1}^{t_2} P_{well,tot}(t)\, dt}_{source} \; . \qquad (2.124)$$

$$\underbrace{\hphantom{E_{pot,tot}(t_2) - E_{pot,tot}(t_1)}}_{accumulation}$$

Note that the potential energy accumulation has simply been expressed as the difference between the values at the begin and end times which illustrates that potential energy is not dependent on the pressure history, i.e. it is path-independent. This is unlike the energy lost by dissipation and gained by work, which are both (pressure) path-dependent as indicated by the integrals which need to be evaluated over the entire time interval. Alternatively, the energy balance per unit time, i.e. the *power balance*, can be expressed as

$$\frac{dE_{pot,tot}(t)}{dt} + \frac{dE_{dis,tot}(t)}{dt} = P_{well,tot}(t) . \qquad (2.125)$$

or, according to Eqs. (2.118), (2.121) and (2.123), as

$$\sum_{i=1}^{n_{gb}} V_i \phi_i c_t p_i(t) \frac{dp_i(t)}{dt} + \sum_{j=1}^{n_{con}} T_j(t) \Delta p_j^2(t) + \sum_{k=1}^{n_{well}} J_{well,k}(t) \left[p_{well,k}(t) - p_{gb,k}(t) \right]^2$$

$$= \sum_{k=1}^{n_{well}} q_{well,k}(t) \times p_{well,k}(t).$$

$$(2.126)$$

Expression (2.126) can also be written in matrix–vector notation as

$$\begin{bmatrix} \mathbf{p}_1 & \mathbf{p}_2 & \mathbf{p}_3 \end{bmatrix} \begin{bmatrix} \mathbf{V}_{11} & \mathbf{0} & \mathbf{0} \\ \mathbf{0} & \mathbf{V}_{22} & \mathbf{0} \\ \mathbf{0} & \mathbf{0} & \mathbf{V}_{33} \end{bmatrix} \begin{bmatrix} \dot{\mathbf{p}}_1 \\ \dot{\mathbf{p}}_2 \\ \dot{\mathbf{p}}_3 \end{bmatrix}$$

$$+ \begin{bmatrix} (\mathbf{p}_1 - \mathbf{p}_{av}) & (\mathbf{p}_2 - \mathbf{p}_{av}) & (\mathbf{p}_3 - \mathbf{p}_{av}) \end{bmatrix} \begin{bmatrix} \mathbf{T}_{11} & \mathbf{T}_{12} & \mathbf{T}_{13} \\ \mathbf{T}_{21} & \mathbf{T}_{22} & \mathbf{T}_{23} \\ \mathbf{T}_{31} & \mathbf{T}_{32} & \mathbf{T}_{33} \end{bmatrix} \begin{bmatrix} \mathbf{p}_1 \\ \mathbf{p}_2 \\ \mathbf{p}_3 \end{bmatrix}$$

$$+ \begin{bmatrix} \mathbf{0} & (\bar{\mathbf{p}}_{well} - \mathbf{p}_2) & (\breve{\mathbf{p}}_{well} - \mathbf{p}_3) \end{bmatrix} \begin{bmatrix} \mathbf{0} & \mathbf{0} & \mathbf{0} \\ \mathbf{0} & \mathbf{J}_2 & \mathbf{0} \\ \mathbf{0} & \mathbf{0} & \mathbf{J}_3 \end{bmatrix} \begin{bmatrix} \mathbf{0} \\ \bar{\mathbf{p}}_{well} - \mathbf{p}_2 \\ \breve{\mathbf{p}}_{well} - \mathbf{p}_3 \end{bmatrix} \qquad (2.127)$$

$$= \begin{bmatrix} \mathbf{0} & \bar{\mathbf{p}}_{well} & \bar{\mathbf{q}}_{well} \end{bmatrix} \begin{bmatrix} \mathbf{0} \\ \breve{\mathbf{q}}_{well} \\ \breve{\mathbf{p}}_{well} \end{bmatrix} ,$$

where we have used the partitioned vectors and matrices as introduced in Sects. 2.2.3 and 2.2.4. The vector $\mathbf{p}_{av}$ represents the time-dependent average reservoir pressure defined as

$$\mathbf{p}_{av}(t) \triangleq \frac{1}{n_{gb}} \sum_{i=1}^{n_{gb}} p_i(t) \begin{bmatrix} 1 \\ 1 \\ \vdots \\ 1 \end{bmatrix}, \tag{2.128}$$

with a number of elements as appropriate to match those of $\mathbf{p}_1$, $\mathbf{p}_2$ and $\mathbf{p}_3$. The equivalence of Eqs. (2.126) and (2.127) can be confirmed by inspection of the matrices $\mathbf{V}$, $\mathbf{T}$ and $\mathbf{J}_3$, and the underlying matrices as defined in Eq. (1.31).

2.2.7.5 Minimum Energy Interpretation*

With the aid of relationships (2.90) and (2.91) we can rewrite the power balance (2.127) as

$$\begin{aligned}
&\begin{bmatrix} \mathbf{p}_1 & \mathbf{p}_2 & \mathbf{p}_3 \end{bmatrix} \begin{bmatrix} \mathbf{V}_{11} & 0 & 0 \\ 0 & \mathbf{V}_{22} & 0 \\ 0 & 0 & \mathbf{V}_{33} \end{bmatrix} \begin{bmatrix} \dot{\mathbf{p}}_1 \\ \dot{\mathbf{p}}_2 \\ \dot{\mathbf{p}}_3 \end{bmatrix} \\
&+ \begin{bmatrix} (\mathbf{p}_1 - \mathbf{p}_{av}) & (\mathbf{p}_2 - \mathbf{p}_{av}) & (\mathbf{p}_3 - \mathbf{p}_{av}) \end{bmatrix} \begin{bmatrix} \mathbf{T}_{11} & \mathbf{T}_{12} & \mathbf{T}_{13} \\ \mathbf{T}_{21} & \mathbf{T}_{22} & \mathbf{T}_{23} \\ \mathbf{T}_{31} & \mathbf{T}_{32} & \mathbf{T}_{33} \end{bmatrix} \begin{bmatrix} \mathbf{p}_1 \\ \mathbf{p}_2 \\ \mathbf{p}_3 \end{bmatrix} \\
&+ \begin{bmatrix} 0 & (\bar{\mathbf{p}}_{well} - \mathbf{p}_2) & (\breve{\mathbf{p}}_{well} - \mathbf{p}_3) \end{bmatrix} \begin{bmatrix} 0 \\ \breve{\mathbf{q}}_{well} \\ \bar{\mathbf{q}}_{well} \end{bmatrix} \\
&= \begin{bmatrix} 0 & \bar{\mathbf{p}}_{well} & \bar{\mathbf{q}}_{well} \end{bmatrix} \begin{bmatrix} 0 \\ \breve{\mathbf{q}}_{well} \\ \breve{\mathbf{p}}_{well} \end{bmatrix},
\end{aligned} \tag{2.129}$$

from which follows

$$\begin{aligned}
&\begin{bmatrix} \mathbf{p}_1 & \mathbf{p}_2 & \mathbf{p}_3 \end{bmatrix} \begin{bmatrix} \mathbf{V}_{11} & 0 & 0 \\ 0 & \mathbf{V}_{22} & 0 \\ 0 & 0 & \mathbf{V}_{33} \end{bmatrix} \begin{bmatrix} \dot{\mathbf{p}}_1 \\ \dot{\mathbf{p}}_2 \\ \dot{\mathbf{p}}_3 \end{bmatrix} \\
&+ \begin{bmatrix} (\mathbf{p}_1 - \mathbf{p}_{av}) & (\mathbf{p}_2 - \mathbf{p}_{av}) & (\mathbf{p}_3 - \mathbf{p}_{av}) \end{bmatrix} \begin{bmatrix} \mathbf{T}_{11} & \mathbf{T}_{12} & \mathbf{T}_{13} \\ \mathbf{T}_{21} & \mathbf{T}_{22} & \mathbf{T}_{23} \\ \mathbf{T}_{31} & \mathbf{T}_{32} & \mathbf{T}_{33} \end{bmatrix} \begin{bmatrix} \mathbf{p}_1 \\ \mathbf{p}_2 \\ \mathbf{p}_3 \end{bmatrix} \\
&= \begin{bmatrix} 0 & \mathbf{p}_2 & \mathbf{p}_3 \end{bmatrix} \begin{bmatrix} 0 \\ \breve{\mathbf{q}}_{well} \\ \bar{\mathbf{q}}_{well} \end{bmatrix}.
\end{aligned} \tag{2.130}$$

Equation (2.130) is again an expression for the power balance in the system, but now expressed solely in terms of the state variables[15] $\mathbf{p}_1$, $\mathbf{p}_2$ and $\mathbf{p}_3$. It can be written compactly as $dE_{sys}(\mathbf{p}_1,\mathbf{p}_2,\mathbf{p}_3)/dt = P_{well}(\mathbf{p}_1,\mathbf{p}_2,\mathbf{p}_3)$, where E_{sys} is the system energy as governed by the state variables proper. Equation (2.130) is a single scalar equation in the n_{gb} state variables, and therefore does not have a unique solution. However, using thermodynamic arguments it can be argued that all natural systems tend to organize themselves in such a way that they minimize the amount of energy required to maintain equilibrium between internal and external forces. In our particular case of a system comprising flow through a porous medium this implies that the values of the state variables $\mathbf{p}_1$, $\mathbf{p}_2$ and $\mathbf{p}_3$ will be such that the power flow through the system becomes minimal i.e. that the first derivatives of P_{well} with respect to the state variables, and therefore also the first derivatives of dE_{sys}/dt with respect to the state variables, become zero. Taking derivatives of Eq. (2.130), setting the results equal to zero and combining them in matrix form results in

$$
\begin{bmatrix} \mathbf{V}_{11} & \mathbf{0} & \mathbf{0} \\ \mathbf{0} & \mathbf{V}_{22} & \mathbf{0} \\ \mathbf{0} & \mathbf{0} & \mathbf{V}_{33} \end{bmatrix} \begin{bmatrix} \dot{\mathbf{p}}_1 \\ \dot{\mathbf{p}}_2 \\ \dot{\mathbf{p}}_3 \end{bmatrix} + \begin{bmatrix} \mathbf{T}_{11} & \mathbf{T}_{12} & \mathbf{T}_{13} \\ \mathbf{T}_{21} & \mathbf{T}_{22} & \mathbf{T}_{23} \\ \mathbf{T}_{31} & \mathbf{T}_{32} & \mathbf{T}_{33} \end{bmatrix} \begin{bmatrix} \mathbf{p}_1 \\ \mathbf{p}_2 \\ \mathbf{p}_3 \end{bmatrix} = \begin{bmatrix} \mathbf{0} \\ \breve{\mathbf{q}}_{well} \\ \bar{\mathbf{q}}_{well} \end{bmatrix} . \qquad (2.131)
$$

Here we made use of the fact that

$$
\frac{d}{d\mathbf{p}}[(\mathbf{p} - \mathbf{p}_{av})\mathbf{Tp}] = 2\mathbf{Tp} - \underbrace{\frac{d\mathbf{p}_{av}}{d\mathbf{p}}}_{\mathbf{I}} \mathbf{Tp} - \underbrace{\mathbf{Tp}_{av}}_{0} = \mathbf{Tp}, \qquad (2.132)
$$

where we used the compact notation $\mathbf{p} = [(\mathbf{p}_1)^T(\mathbf{p}_2)^T(\mathbf{p}_3)^T]^T$, etc., as introduced in Sects. 2.2.3, leaving out the superscripted stars for clarity. Using Eq. (2.90), we can finally rewrite Eq. (2.131) as

$$
\begin{bmatrix} \mathbf{V}_{11} & \mathbf{0} & \mathbf{0} \\ \mathbf{0} & \mathbf{V}_{22} & \mathbf{0} \\ \mathbf{0} & \mathbf{0} & \mathbf{V}_{33} \end{bmatrix} \begin{bmatrix} \dot{\mathbf{p}}_1 \\ \dot{\mathbf{p}}_2 \\ \dot{\mathbf{p}}_3 \end{bmatrix} + \begin{bmatrix} \mathbf{T}_{11} & \mathbf{T}_{12} & \mathbf{T}_{13} \\ \mathbf{T}_{21} & \mathbf{T}_{22} & \mathbf{T}_{23} \\ \mathbf{T}_{31} & \mathbf{T}_{32} & \mathbf{T}_{33} + \mathbf{J}_3 \end{bmatrix} \begin{bmatrix} \mathbf{p}_1 \\ \mathbf{p}_2 \\ \mathbf{p}_3 \end{bmatrix}
$$
$$
= \begin{bmatrix} \mathbf{0} & \mathbf{0} & \mathbf{0} \\ \mathbf{0} & \mathbf{I} & \mathbf{0} \\ \mathbf{0} & \mathbf{0} & \mathbf{J}_3 \end{bmatrix} \begin{bmatrix} \mathbf{0} \\ \breve{\mathbf{q}}_{well} \\ \breve{\mathbf{p}}_{well} \end{bmatrix} . \qquad (2.133)
$$

Equation (2.133) is completely identical to system Eq. (2.89) which was derived from balance equations for mass and momentum only. The alternative way to derive the system equations using the concept of energy, as described for porous-medium flow in this section, is frequently used in the fields of theoretical and applied mechanics; see e.g. Langhaar (1962) and Lanczos (1970). Closely related are other *energy methods* also known as *variational methods*, which are used to

---

[15] In comparison, Eq. (2.127) was also a function of the well-bore pressures $\breve{\mathbf{p}}_{well}$ and $\bar{\mathbf{p}}_{well}$.

compute approximate solutions for applied mechanics problems in complex-shaped domains. In particular, they often form the basis to derive numerical approximations using the finite-element method; see e.g. Zienckiewicz and Taylor (1989). Direct use of energy methods in porous-media flow does not seem to have an advantage over the conventional direct methods, and has therefore scarcely been described in the literature. An exception is the paper by Karney and Seneviratne (1991) who propose to use energy concepts for adaptive time step control in numerical simulation.

2.3 Two-Phase Flow

2.3.1 System Equations

2.3.1.1 Nonlinear Equations

As a next step we consider a simplified description of two-phase flow of oil and water, as derived in some detail in Sect. 1.4. We start from Eq. (1.129),

$$\underbrace{\begin{bmatrix} \mathbf{V}_{wp}(\mathbf{s}) & \mathbf{V}_{ws} \\ \mathbf{V}_{op}(\mathbf{s}) & \mathbf{V}_{os} \end{bmatrix}}_{\mathbf{V}} \begin{bmatrix} \dot{\mathbf{p}} \\ \dot{\mathbf{s}} \end{bmatrix} + \underbrace{\begin{bmatrix} \mathbf{T}_w(\mathbf{s}) & 0 \\ \mathbf{T}_o(\mathbf{s}) & 0 \end{bmatrix}}_{\mathbf{T}} \begin{bmatrix} \mathbf{p} \\ \mathbf{s} \end{bmatrix} = \underbrace{\begin{bmatrix} \mathbf{F}_w(\mathbf{s}) \\ \mathbf{F}_o(\mathbf{s}) \end{bmatrix}}_{\mathbf{F}} \mathbf{q}_{well,t} \,, \qquad (2.134)$$

where $\mathbf{p}$ and $\mathbf{s}$ are vectors of pressures p_o and water saturations S_w in the grid-block centers, $\mathbf{V}$ is the accumulation matrix (with entries that are functions of the porosity ϕ, and the oil, water and rock compressibilities c_o, c_w and c_r), $\mathbf{T}$ is the transmissibility matrix (with entries that are functions of the rock permeabilities k, the oil and water relative permeabilities k_{ro} and k_{rw} and the oil and water viscosities μ_o and μ_w), $\mathbf{F}$ is the fractional-flow matrix (with entries that have functional dependencies similar to those of $\mathbf{T}$), and $\mathbf{q}_{well,t}$ is a vector of total well flow rates with non-zero values in those elements that correspond to grid blocks penetrated by a well. The nonlinearities in Eq. (2.134) result from various sources; see also Sect. 1.4.10. If the oil and water phases have different compressibilities, the accumulation terms $\mathbf{V}_{wp}(\mathbf{s})$ and $\mathbf{V}_{op}(\mathbf{s})$ are (weak) functions of the saturations because the liquid compressibility is a saturation-weighted average of the oil and water compressibilities. Moreover, the porosity and compressibility values in these terms may be weak functions of pressure, but we do not take this effect into account in the examples in this text. The transmissibility terms $\mathbf{T}_w(\mathbf{s})$ and $\mathbf{T}_o(\mathbf{s})$ are much stronger functions of the saturations, because the relative permeabilities for oil and water are strongly saturation-dependent. The viscosities may also be weakly pressure-dependent, but, yet again, the pressure dependency is disregarded in the examples. Finally, matrices $\mathbf{F}_o(\mathbf{s})$ and $F_w(\mathbf{s})$ contain saturation-dependent terms that relate the oil and water flow rates in the wells to the total flow rates.

2.3.1.2 Well Model

In practice the source terms are often not the flow rates in the wells but rather the pressures. This can be accounted for by rewriting Eq. (2.134) in partitioned form as

$$
\begin{bmatrix}
\mathbf{V}_{wp,11} & 0 & 0 & \mathbf{V}_{ws,11} & 0 & 0 \\
0 & \mathbf{V}_{wp,22} & 0 & 0 & \mathbf{V}_{ws,22} & 0 \\
0 & 0 & \mathbf{V}_{wp,33} & 0 & 0 & \mathbf{V}_{ws,33} \\
\mathbf{V}_{op,11} & 0 & 0 & \mathbf{V}_{os,11} & 0 & 0 \\
0 & \mathbf{V}_{op,22} & 0 & 0 & \mathbf{V}_{os,22} & 0 \\
0 & 0 & \mathbf{V}_{op,33} & 0 & 0 & \mathbf{V}_{os,33}
\end{bmatrix}
\begin{bmatrix}
\dot{\mathbf{p}}_1 \\ \dot{\mathbf{p}}_2 \\ \dot{\mathbf{p}}_3 \\ \dot{\mathbf{s}}_1 \\ \dot{\mathbf{s}}_2 \\ \dot{\mathbf{s}}_3
\end{bmatrix} +
$$

$$
\begin{bmatrix}
\mathbf{T}_{w,11} & \mathbf{T}_{w,12} & \mathbf{T}_{w,13} & 0 & 0 & 0 \\
\mathbf{T}_{w,21} & \mathbf{T}_{w,22} & \mathbf{T}_{w,23} & 0 & 0 & 0 \\
\mathbf{T}_{w,31} & \mathbf{T}_{w,32} & \mathbf{T}_{w,33} & 0 & 0 & 0 \\
\mathbf{T}_{o,11} & \mathbf{T}_{o,12} & \mathbf{T}_{o,13} & 0 & 0 & 0 \\
\mathbf{T}_{o,21} & \mathbf{T}_{o,22} & \mathbf{T}_{o,23} & 0 & 0 & 0 \\
\mathbf{T}_{o,31} & \mathbf{T}_{o,32} & \mathbf{T}_{o,33} & 0 & 0 & 0
\end{bmatrix}
\begin{bmatrix}
\mathbf{p}_1 \\ \mathbf{p}_2 \\ \mathbf{p}_3 \\ \mathbf{s}_1 \\ \mathbf{s}_2 \\ \mathbf{s}_3
\end{bmatrix} =
\begin{bmatrix}
0 & 0 & 0 \\
0 & \mathbf{F}_{w,22} & 0 \\
0 & 0 & \mathbf{F}_{w,33} \\
0 & 0 & 0 \\
0 & \mathbf{F}_{o,22} & 0 \\
0 & 0 & \mathbf{F}_{o,33}
\end{bmatrix}
\begin{bmatrix}
0 \\
\breve{\mathbf{q}}_{well,t} \\
\mathbf{J}_3\left(\breve{\mathbf{p}}_{well} - \mathbf{p}_3\right)
\end{bmatrix}.
$$

$$(2.135)$$

Here, the elements of vector $\mathbf{p}_1$ are the pressures in those grid blocks that are not penetrated by a well. The elements of $\mathbf{p}_2$ are the pressures in the blocks where the source terms are prescribed total well flow rates $\breve{\mathbf{q}}_{well,t}$, and those of $\mathbf{p}_3$ are the pressures in the blocks where the source terms are obtained through prescription of the bottom-hole pressures $\breve{\mathbf{p}}_{well}$ with the aid of a diagonal matrix of well indices $\mathbf{J}_3$. To compute the oil and water flow rates in the wells with prescribed pressures we use the well model

$$
\begin{bmatrix}
\bar{\mathbf{q}}_{well,w} \\
\bar{\mathbf{q}}_{well,o}
\end{bmatrix} =
\begin{bmatrix}
\mathbf{F}_{w,33} \\
\mathbf{F}_{o,33}
\end{bmatrix}
\mathbf{J}_3\left(\breve{\mathbf{p}}_{well} - \mathbf{p}_3\right). \tag{2.136}
$$

To compute the bottom-hole pressures $\bar{\mathbf{p}}_{well}$ in the wells with prescribed total flow rates we need an additional diagonal matrix $\mathbf{J}_{q,2}$ of well indices such that

$$
\breve{\mathbf{q}}_{well,t} = \mathbf{J}_2(\bar{\mathbf{p}}_{well} - \mathbf{p}_2), \tag{2.137}
$$

from which we obtain

$$
\bar{\mathbf{p}}_{well} = \mathbf{J}_2^{-1}\breve{\mathbf{q}}_{well,t} - \mathbf{p}_2. \tag{2.138}
$$

2.3.1.3 State-Space Form

To bring these equations in state-space form we define the state vector $\mathbf{x}$, input vector $\mathbf{u}$ and output vector $\mathbf{y}$ as

$$\mathbf{u} \triangleq \begin{bmatrix} \breve{\mathbf{q}}_{well,t} \\ \breve{\mathbf{p}}_{well} \end{bmatrix}, \tag{2.139}$$

$$\mathbf{x} \triangleq \begin{bmatrix} \mathbf{p} \\ \mathbf{s} \end{bmatrix} = \begin{bmatrix} p_1 \\ p_2 \\ p_3 \\ s_1 \\ s_2 \\ s_3 \end{bmatrix}, \tag{2.140}$$

$$\mathbf{y} \triangleq \begin{bmatrix} \bar{\mathbf{p}}_{well} \\ \bar{\mathbf{q}}_{well,w} \\ \bar{\mathbf{q}}_{well,o} \end{bmatrix}. \tag{2.141}$$

Equations (2.135), (2.136) and (2.138) can then be rewritten in nonlinear state-space form

$$\dot{\mathbf{x}} = \mathbf{f}(\mathbf{u}, \mathbf{x}), \tag{2.142}$$

$$\mathbf{y} = \mathbf{h}(\mathbf{u}, \mathbf{x}), \tag{2.143}$$

where the functions $\mathbf{f}$ and $\mathbf{h}$ are defined as

$$\mathbf{f} \triangleq \mathbf{A}\mathbf{x} + \mathbf{B}\mathbf{u}, \tag{2.144}$$

$$\mathbf{h} \triangleq \mathbf{C}\mathbf{x} + \mathbf{D}\mathbf{u}, \tag{2.145}$$

with state-dependent secant matrices $\mathbf{A}(\mathbf{x})$, $\mathbf{B}(\mathbf{x})$, $\mathbf{C}(\mathbf{x})$ and $\mathbf{D}(\mathbf{x})$ given by

$$\mathbf{A} \triangleq -\mathbf{V}^{-1} \underbrace{\left[\begin{array}{ccc|ccc} \mathbf{T}_{w,11} & \mathbf{T}_{w,12} & \mathbf{T}_{w,13} & \mathbf{0} & \mathbf{0} & \mathbf{0} \\ \mathbf{T}_{w,21} & \mathbf{T}_{w,22} & \mathbf{T}_{w,23} & \mathbf{0} & \mathbf{0} & \mathbf{0} \\ \mathbf{T}_{w,31} & \mathbf{T}_{w,32} & \mathbf{T}_{w,33} + \mathbf{F}_{w,33}\mathbf{J}_3 & \mathbf{0} & \mathbf{0} & \mathbf{0} \\ \hline \mathbf{T}_{o,11} & \mathbf{T}_{o,12} & \mathbf{T}_{o,13} & \mathbf{0} & \mathbf{0} & \mathbf{0} \\ \mathbf{T}_{o,21} & \mathbf{T}_{o,22} & \mathbf{T}_{o,23} & \mathbf{0} & \mathbf{0} & \mathbf{0} \\ \mathbf{T}_{o,31} & \mathbf{T}_{o,32} & \mathbf{T}_{o,33} + \mathbf{F}_{o,33}\mathbf{J}_3 & \mathbf{0} & \mathbf{0} & \mathbf{0} \end{array}\right]}_{-\hat{\mathbf{A}}} \tag{2.146}$$

$$B \triangleq V^{-1} \underbrace{\begin{bmatrix} 0 & 0 \\ F_{w,22} & 0 \\ 0 & F_{w,33} J_3 \\ 0 & 0 \\ F_{o,22} & 0 \\ 0 & F_{o,33} J_3 \end{bmatrix}}_{\hat{B}}, \tag{2.147}$$

$$C \triangleq \begin{bmatrix} 0 & I & 0 & | & 0 & 0 & 0 \\ 0 & 0 & -F_{w,33} J_3 & | & 0 & 0 & 0 \\ 0 & 0 & -F_{o,33} J_3 & | & 0 & 0 & 0 \end{bmatrix}, \tag{2.148}$$

$$D \triangleq \begin{bmatrix} J_2^{-1} & 0 \\ 0 & F_{w,33} J_3 \\ 0 & F_{o,33} J_3 \end{bmatrix}. \tag{2.149}$$

We note that the explicit representations (2.142) and (2.143) are primarily of theoretical interest because they allow direct application of concepts from systems-and-control theory. For computational purposes it is usually required to express the system equations in fully-implicit (residual) state-space form

$$g(u, x, \dot{x}) = \hat{E}\dot{x} - \hat{A}x - \hat{B}u = 0, \tag{2.150}$$

where $\hat{E} = V$ and where $\hat{A}$ and $\hat{B}$ are have been defined in Eqs. (2.146) and (2.147). The computation of the inverse of V as required in the explicit form can be performed efficiently by using the analytical expression for the inverse of a 2×2 block matrix with diagonal blocks of equal size[16]:

$$\begin{bmatrix} V_{wp} & V_{ws} \\ V_{op} & V_{os} \end{bmatrix}^{-1} = \begin{bmatrix} \tilde{V}^{-1} V_{os} & -\tilde{V}^{-1} V_{ws} \\ -\tilde{V}^{-1} V_{op} & \tilde{V}^{-1} V_{wp} \end{bmatrix}, \tag{2.151}$$

where

$$\tilde{V} = V_{wp} V_{os} - V_{ws} V_{op}. \tag{2.152}$$

Because the four sub-matrices of V are diagonal, $\tilde{V}$ and the four sub-matrices of V^{-1} are also diagonal. Moreover, the inverse $\tilde{V}^{-1}$ can be obtained simply by

---

[16] The general expression for the inverse of a 2×2 block matrix is given by $\begin{bmatrix} V_{11} & V_{12} \\ V_{21} & V_{22} \end{bmatrix}^{-1} = \begin{bmatrix} \tilde{V}_1^{-1} & -V_{11}^{-1} V_{12} \tilde{V}_2^{-1} \\ -V_{22}^{-1} V_{21} \tilde{V}_1^{-1} & \tilde{V}_2^{-1} \end{bmatrix}$, where $\tilde{V}_1 = V_{11} - V_{12} V_{22}^{-1} V_{21}$ and $\tilde{V}_2 = V_{22} - V_{11}^{-1} V_{12}$ are the *Schur complements* of V_{11} and V_{22} respectively; see e.g. Friedland (1986), pp. 479–481. Using the property that equally sized diagonal matrices commute, we can derive Eq. (2.151) from this more general expression.

taking the reciprocals of the diagonal elements. However, we re-emphasize that there is no need to perform the inverse operation if the equations serve as a basis for computation, and that the explicit state-space form (2.142) is only required for analysis of the system-theoretical properties of the equations.

2.3.1.4 Extended Output vector

In the formulation discussed so far we considered system outputs in the sense of response signals, and we tacitly assumed that all system inputs were known. However, in reality, both inputs and outputs have to be measured and we can therefore also define an extended output vector that contains all measured signals. For example it may be required to know the oil and water flow rates in those wells were the total flow rates have been prescribed, which leads to

$$\begin{bmatrix} \tilde{\mathbf{q}}_{well,w} \\ \tilde{\mathbf{q}}_{well,o} \end{bmatrix} = \begin{bmatrix} \mathbf{F}_{w,22} \\ \mathbf{F}_{o,22} \end{bmatrix} \tilde{\mathbf{q}}_{well,t} ,$$ (2.153)

where we have added tildes to indicate that the variables are measurements rather than real prescribed variables. Moreover, we may want to include measurements of the prescribed pressure $\tilde{\mathbf{p}}_{well}$ in the output. The extended output vector then becomes

$$\mathbf{y} \triangleq \begin{bmatrix} \bar{\mathbf{p}}_{well} \\ \bar{\mathbf{q}}_{well,w} \\ \bar{\mathbf{q}}_{well,o} \\ \hdashline \tilde{\mathbf{p}}_{well} \\ \tilde{\mathbf{q}}_{well,w} \\ \tilde{\mathbf{q}}_{well,o} \end{bmatrix} ,$$ (2.154)

where the elements above and below the dotted line represent measurements related to output and input variables respectively.[17] In this case the matrices $\mathbf{C}$ and $\mathbf{D}$ can be expressed as

$$\mathbf{C} \triangleq \begin{bmatrix} \mathbf{0} & \mathbf{I} & \mathbf{0} & \mathbf{0} & \mathbf{0} & \mathbf{0} \\ \mathbf{0} & \mathbf{0} & -\mathbf{F}_{w,33}\,\mathbf{J}_3 & \mathbf{0} & \mathbf{0} & \mathbf{0} \\ \mathbf{0} & \mathbf{0} & -\mathbf{F}_{o,33}\,\mathbf{J}_3 & \mathbf{0} & \mathbf{0} & \mathbf{0} \\ \mathbf{0} & \mathbf{0} & \mathbf{0} & \mathbf{0} & \mathbf{0} & \mathbf{0} \\ \mathbf{0} & \mathbf{0} & \mathbf{0} & \mathbf{0} & \mathbf{0} & \mathbf{0} \\ \mathbf{0} & \mathbf{0} & \mathbf{0} & \mathbf{0} & \mathbf{0} & \mathbf{0} \end{bmatrix} ,$$ (2.155)

---

[17] This distinction is not very clear cut. For example to compute the oil and water 'input' rates, we make use of the fractional flows around the wells which are a direct function of the saturations, i.e. of state variables. In this sense the rates also contain indirect output information on the saturations around the wells.

$$\mathbf{D} \triangleq \begin{bmatrix} \mathbf{J}_2^{-1} & \mathbf{0} \\ \mathbf{0} & \mathbf{F}_{w,33}\,\mathbf{J}_3 \\ \mathbf{0} & \mathbf{F}_{o,33}\,\mathbf{J}_3 \\ \hline \mathbf{0} & \mathbf{I} \\ \mathbf{F}_{w,22} & \mathbf{0} \\ \mathbf{F}_{o,22} & \mathbf{0} \end{bmatrix}. \tag{2.156}$$

2.3.2 Well Operating Constraints

During the operation of an oil field it often occurs that wells or groups of wells are operated on liquid constraints because the surface facilities are not capable of processing more than a certain throughput of gas, oil and water. Water injection wells are often operated on pressure constraints to avoid or limit fracturing of the formation around the wells. Production wells are often constrained to operate at a tubing-head pressure above a certain minimum, as determined by the working pressure of the first separator plus some additional pressure to displace the fluids through the flow line to that separator. Moreover, during the producing life of a field the well operating constraints may change because of changes in reservoir pressure and well-bore stream composition. In practice the control of tubing head pressures or phase rates is often done indirectly, through adjusting valve settings or changing out chokes and monitoring the resulting pressure of flow rate response. Methods for well control vary drastically. At the high end we find sophisticated remotely controlled valves with remotely observed pressure gauges and multi-phase flow meters. At the low end we have manual change out of chokes, and infrequent, say monthly, observations of well head pressures and measurements of the well phase rates by temporarily re-routing the well through a test separator. In reservoir simulation we can prescribe pressures or flow rates, but, in addition we may specify constraints in the form of maximum or minimum values for pressures and flow rates in wells or groups of wells. During the simulation the conditions may change such that a well changes from being operated at a prescribed rate to being operated at a prescribed pressure or vice versa. For example if a production well is operated at a prescribed total liquid rate, reservoir depletion may cause the bottom-hole pressure required to maintain this flow rate to gradually drop until it reaches the minimum pressure required to lift the well-bore fluid to surface, i.e. until it reaches its minimum pressure constraint. From that moment on the well needs to be operated at a prescribed bottom-hole pressure. Most reservoir simulators therefore allow the user to define minimum and maximum constraints for pressures and phase rates and automatically determine the *most constraining constraint* at any moment in time during the simulation. Examples of operating constraints as implemented in reservoir simulation will be discussed in Chap. 3.

2.3.3   Computational Aspects

In this section we discuss some general computational aspects of the numerical implementation of the two-phase system equations.

- Most (sub-)matrices considered so far are *sparse*: the accumulation sub-matrices are diagonal, the transmissibility sub-matrices are penta-diagonally banded with two sub diagonals, and the fractional-flow and well index sub-matrices are sparse diagonal. Most of the matrix elements are therefore equal to zero, and this property may be used to significantly reduce computer memory usage.
- As mentioned before, the re-ordering of vector and matrix elements with permutation matrices as used in Sects. 2.2.3 to Sects. 2.2.6 is not essential in a numerical implementation. There is no computational need to e.g. re-group state or input vector elements; it is merely a convenient notation. In a numerical implementation we may simply use (sparse) matrices with elements that correspond to the relevant state or input variables at the appropriate locations.
- Computation of an element of a transmissibility sub-matrix corresponding to a specific grid block involves computing the transmissibilities for flow to or from the four neighboring grid blocks. Therefore, assembling the transmissibility (sub-)matrices requires knowledge of the *connectivities* of the grid blocks. This knowledge is often administered with the aid of a *connectivity* table, an $n_{con} \times 2$ matrix of which each row corresponds to a connectivity between a pair of grid blocks with grid-block numbers stored in the first and second column positions. For a rectangular model with $n_x \times n_y$ grid blocks, we have

$$n_{con} = (n_x - 1)n_y + (n_y - 1)n_x, \qquad (2.157)$$

and for the 2 × 3 reservoir used in Examples 1 and 2 the 7 × 2 connectivity table $\mathbf{L}_{con}$ is given by (see also Table 1.2)

$$\mathbf{L}_{con} = \begin{bmatrix} 1 & 2 \\ 1 & 4 \\ 2 & 3 \\ 2 & 5 \\ 3 & 6 \\ 4 & 5 \\ 5 & 6 \end{bmatrix}, \qquad (2.158)$$

- The elements in the two-phase state vector $\mathbf{x} = [\mathbf{p}^T \mathbf{s}^T]^T$ have different physical dimensions and strongly varying magnitudes. If we express the pressures in Pa, they are in the order of 10^6–10^7, whereas the saturation values remain, by definition, between 0 and 1. As a result the elements of the transmissibility matrix $\mathbf{T}$, and therefore of the system matrix $\mathbf{A}$, will also have strongly varying magnitude. This may influence the accuracy of the result when solving a system of equations $\mathbf{A}\,\mathbf{x} = \mathbf{b}$, as will be required in Chap. 3 to simulate the response of the system[18]. The reason for the inaccuracy is in the finite precision representation of the matrix elements in any numerical implementation. We may avoid this problem by scaling the elements of $\mathbf{x}$ such that the difference in magnitude between the pressure and saturation values becomes much smaller. This can be done by dividing the first n_{gb} columns of $\mathbf{A}$, which multiply the first n_{gb} 'pressure' elements of $\mathbf{x}$, by a factor

$$f_{scal} = \max(\mathbf{p}) \,. \qquad (2.159)$$

and multiply the corresponding elements of $\mathbf{x}$ with f_{scal} after the equations have been solved. In addition, we may also scale the elements of the right-hand side $\mathbf{b}$ by dividing the first n_{gb} rows of $\mathbf{b}$ and the corresponding rows of $\mathbf{A}$ by f_{scal}.

- In an injection well we have $\mathbf{q}_t = \mathbf{q}_w$, and we expect that soon after injection has started the fractional flows for water and oil close to the well will approach one and zero respectively. However, before injection starts, the initial condition for the saturation is usually equal to the connate-water saturation, which means that the fractional flows for water and oil are zero and one respectively. In theory, it would then be impossible to ever inject water. This paradox is usually circumvented by simply specifying a fractional flow equal to one for every injection well.

2.3.4 Lift Tables*

Until now, we have considered prescribed pressures in the form of flowing bottom-hole pressures p_{well}. In most cases, however, it is not the bottom-hole pressure that is controlled but the pressure at the top of the well, which is usually referred to as the flowing tubing-head pressure, p_{tf}. The difference in pressure between top and bottom of a well is governed by the multi-phase flow behavior of the well-bore fluids. Various techniques have been developed to compute well-bore pressure drops, ranging from empirical correlations to complex mechanistic models; see e.g. Brill and Mukherjee 1999. Typically, the tubing-head pressure can be computed for given fluid properties, well-bore geometry, oil, gas and water flow rates, and bottom-hole pressure. Conversely, the bottom-hole pressure may be computed for a given tubing-head pressure. The computation is performed with the aid of a *well-bore simulator* that numerically integrates a one-dimensional averaged version of the multi-phase flow equations along the well bore. Especially in the case of complex mechanistic multi-phase flow models these computations may be too

---

[18] Here, $\mathbf{b}$ is an arbitrary right-hand side.

Fig. 2.2 Schematic representation of a smart horizontal well equipped with three ICVs. These allow for individual control of the inflow from the reservoir, through the perforations and the ICVs, into the tubing. The packers form flow barriers in the annular space between the casing and the tubing

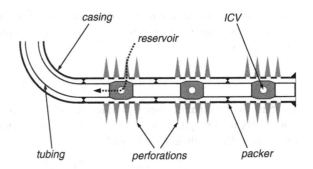

time consuming to perform every time step of the reservoir simulator. An alternative approach is then to perform a large number of well-bore flow simulations up-front to generate a multi-dimensional table, known as a lift table or *flow performance* table, which can be used as a look-up table by the reservoir simulator. Usually the four entries for a lift table are the tubing-head pressure, and the oil, gas and water rates, all expressed at standard conditions.[19] Typically each of the entries is described with a small number of points, say 5 in which case the table has $5^4 = 625$ points that correspond to the same number of bottom-hole pressures. For intermediate values of the entries a linear or higher-order interpolation is used to compute the corresponding bottom-hole pressure, which is much faster than performing a full well-bore flow simulation. Sometimes a higher number of points is needed, at the cost of a longer pre-processing time, e.g. to prevent convergence problems during the numerical solution of the reservoir equations.

2.3.5 Control Valves*

In addition to specifying inputs in the form of prescribed well flow rates or pressures, either down hole or at surface, it is also possible to prescribe the opening of control valves. In particular the use of *interval control valves* (ICVs) is becoming increasingly popular. ICVs are mounted in an inner tube (the *tubing*) inside an outer tube (the *casing*) which penetrates the reservoir; see Fig. 2.2. The role of these valves is to control the inflow from individual reservoir compartments, allowing one, e.g., to shut-in a zone that experiences a too-high water production.

---

[19] Even if the reservoir is above bubble point such that it contains only oil and water and no free gas, the flow in the well bore will be three-phase because associated gas   will be released from the oil as the well-bore pressure decreases at increasing elevations above the reservoir. Alternatively, the lift table entries can be chosen as tubing-head pressure, oil rate, gas-oil ratio, and water-cut. Whatever the choice of the table entries, it is assumed that the fluid properties at standard conditions and the well-bore geometry do not change during the reservoir simulation.

The usual way to represent the control action of such a valve is to specify a dimensionless valve opening $0 \leq \alpha \leq 1$ and modify the well inflow Eq. (2.86) according to

$$q = \alpha J_{well}\left(\breve{p}_{well} - p\right).$$
(2.160)

To compute the oil and water flow rates in wells with prescribed valve settings we can modify the two-phase well model (2.136). Introducing a subscript 4 to indicate wells controlled with ICVs this leads to

$$\begin{bmatrix} \bar{\mathbf{q}}_{well,w} \\ \bar{\mathbf{q}}_{well,o} \end{bmatrix} = \begin{bmatrix} \mathbf{F}_{w,44} \\ \mathbf{F}_{o,44} \end{bmatrix} \mathbf{J}_\alpha \, \boldsymbol{\alpha},$$
(2.161)

where α is a vector of valve settings, and $\mathbf{J}_\alpha$ is a modified well index matrix defined as

$$\mathbf{J}_\alpha \triangleq \begin{bmatrix} J_{well,4,1}\left(p_{well} - p_{4,1}\right) & 0 & \cdots & 0 \\ 0 & J_{well,4,2}\left(p_{well} - p_{4,2}\right) & \cdots & 0 \\ \vdots & \vdots & \ddots & \vdots \\ 0 & 0 & \cdots & J_{well,4,m}\left(p_{well} - p_{4,m}\right) \end{bmatrix}.$$
(2.162)

Note that $\mathbf{J}_\alpha$ is a function of both $\mathbf{p}$ and $\mathbf{s}$. If we define the input and state vectors,

$$\mathbf{u} \triangleq \boldsymbol{\alpha}$$
(2.163)

$$\mathbf{x} \triangleq \begin{bmatrix} \mathbf{p} \\ \mathbf{s} \end{bmatrix} = \begin{bmatrix} \mathbf{p}_1 \\ \mathbf{p}_4 \\ \mathbf{s}_1 \\ \mathbf{s}_4 \end{bmatrix},$$
(2.164)

where subscripts 1 and 4 refer to grid blocks without wells and grid blocks containing an ICV respectively, and where it has been assumed that all ICVs operate at the same, fixed, well pressure p_{well}, the matrices $\mathbf{A}$ and $\mathbf{B}$ become (c.f. Eqs. (2.146) and (2.147)):

$$\mathbf{A} \triangleq -\mathbf{V}^{-1} \begin{bmatrix} \mathbf{T}_{w,11} & \mathbf{T}_{w,14} & \mathbf{0} & \mathbf{0} \\ \mathbf{T}_{w,41} & \mathbf{T}_{w,44} & \mathbf{0} & \mathbf{0} \\ \mathbf{T}_{o,11} & \mathbf{T}_{o,14} & \mathbf{0} & \mathbf{0} \\ \mathbf{T}_{o,41} & \mathbf{T}_{o,44} & \mathbf{0} & \mathbf{0} \end{bmatrix},$$
(2.165)

$$\mathbf{B} \triangleq \mathbf{V}^{-1} \begin{bmatrix} 0 \\ \dfrac{\mathbf{F}_{w,44}\mathbf{J}_\alpha}{0} \\ \mathbf{F}_{o,44}\mathbf{J}_\alpha \end{bmatrix}. \qquad (2.166)$$

In this form the system equations $\dot{\mathbf{x}} = \mathbf{A}(\mathbf{x})\mathbf{x} + \mathbf{B}(\mathbf{x})\mathbf{u}$ are still control-affine.[20] If we introduce the option to control both the bottom-hole pressures and the valve settings, the inputs become nonlinear in $\mathbf{u}$ and the control-affine property is lost.

2.3.6 Streamlines*

As discussed in Chap. 1, the governing equations for flow through porous media consist of a mass-balance equation in combination with Darcy's law which describes the relationship between spatial pressure gradients and fluid velocities. After spatial discretization, Darcy's law can be interpreted as an equation relating pressure differences between adjacent grid blocks to the Darcy velocities (volumetric fluxes) at the corresponding grid-block boundaries. This discrete form of Darcy's law can be expressed as

$$v_t = \mathbf{Sp}, \qquad (2.167)$$

where $\mathbf{p}$ is an $n_{gb} \times 1$ vector of pressures at the grid-block centers with n_{gb} the number of grid blocks, v_t is an $n_{con} \times 1$ vector of total Darcy velocities at the grid-block boundaries with n_{con} the number of connectivities, i.e. the number of grid-block boundaries, and $\mathbf{S}$ is an $n_{con} \times n_{gb}$ matrix of transmissibility coefficients. Expressions for the elements of $\mathbf{S}$ are given in detail in Eq. (1.144) in Sect. 1.4.12. Given the velocity vector v_t, we can now simply visualize the trajectory of a fluid particle starting from its entrance into the reservoir at an injection well, all the way until it leaves again via a producer. These trajectories are known as *streamlines* and they can be computed using a procedure due to Pollock (1988). Consider a two-dimensional reservoir model with total Darcy velocities at the grid-block boundaries given by the vector v_t. The corresponding total interstitial velocities are then given by

$$\tilde{v}_t = \frac{v_t}{\phi}. \qquad (2.168)$$

Assuming a linear change in velocities in the x and y directions we can define the velocity gradients g_x and g_y for a single grid block as

$$g_x = \frac{\tilde{v}_{x_0+\Delta x} - \tilde{v}_{x_0}}{\Delta x}, \qquad (2.169)$$

---

[20] See the footnote on page 46.

Fig. 2.3 Grid block with
velocity vectors at the
boundaries and a streamline
from entrance point (x_i, y_i) to
exit point (x_e, y_e)

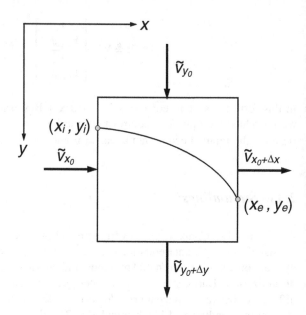

$$g_y = \frac{\tilde{v}_{y_0+\Delta y} - \tilde{v}_{y_0}}{\Delta y}, \qquad (2.170)$$

where Δx and Δy are the grid-block dimensions, where we dropped the subscript
t from the velocities for clarity, and where we used subscripts x_0, y_0, $x_0 + \Delta x$, and
$y_0 + \Delta y$ to indicate the four relevant elements out of the m elements of $\tilde{v}_t$; see
Fig. 2.3. In case of positive velocity components, a fluid particle will enter the
grid block either at the left or at the top. We indicate the location of the entrance
point as (x_i, y_i), where it should be understood that either $x_i = x_0$ or $y_i = y_0$ (or
both, in the special case that the particle enters at the corner). The particle will now
travel along a curved path until it reaches the exit point (x_e, y_e), and its velocity at
an arbitrary point (x, y) inside the grid block has components

$$\tilde{v}_x = \tilde{v}_{x_0} + g_x(x - x_0), \qquad (2.171)$$

$$\tilde{v}_y = \tilde{v}_{y_0} + g_y(y - y_0). \qquad (2.172)$$

Note that at the exit point either $x_e = x_0 + \Delta x$ or $y_e = y_0 + \Delta x$, except when the
particle leaves at the corner. Because we have, by definition,

$$\tilde{v}_x = \frac{dx}{dt}, \qquad (2.173)$$

it follows that

$$dt = \frac{1}{\tilde{v}_x} dx, \qquad (2.174)$$

which can be integrated to obtain the time $\Delta\tau$ to travel the distance $x_e - x_i$:

$$\int\limits_0^{\Delta\tau} dt = \int\limits_{x_i}^{x_e} \frac{1}{\tilde{v}_x} dx = \int\limits_{x_i}^{x_e} \frac{1}{\tilde{v}_{x_0} + g_x(x - x_0)} dx = \frac{1}{g_x} \ln[\tilde{v}_{x_0} + g_x(x - x_0)]|_{x_i}^{x_e}, \quad (2.175)$$

from which we obtain

$$\Delta\tau = \frac{1}{g_x} \ln\left[\frac{\tilde{v}_{x_0} + g_x(x_e - x_0)}{\tilde{v}_{x_0} + g_x(x_i - x_0)}\right]. \quad (2.176)$$

The travel time in the y direction will of course be identical and we can therefore also write

$$\Delta\tau = \frac{1}{g_y} \ln\left[\frac{\tilde{v}_{y_0} + g_y(y_e - y_0)}{\tilde{v}_{y_0} + g_y(y_i - y_0)}\right]. \quad (2.177)$$

We do not know in advance whether the particle will exit at the right or at the bottom of the grid block, but we do know that it must be one of the two (or both, in case of an exit at the corner). To determine the correct exit boundary we should first compute the travel times from Eqs. (2.176) and (2.177) using $x_e - x_0 = \Delta x$ and $y_e - y_0 = \Delta y$ respectively, i.e. for the maximum possible travel distance:

$$\Delta\tau_x = \frac{1}{g_x} \ln\left[\frac{\tilde{v}_{x_0} + g_x\Delta x}{\tilde{v}_{x_0} + g_x(x_i - x_0)}\right], \quad (2.178)$$

$$\Delta\tau_y = \frac{1}{g_y} \ln\left[\frac{\tilde{v}_{y_0} + g_y\Delta y}{\tilde{v}_{y_0} + g_y(y_i - y_0)}\right], \quad (2.179)$$

and the determine the correct grid-block travel time as

$$\Delta\tau = \min(\Delta\tau_x, \Delta\tau_y). \quad (2.180)$$

In the case that $\Delta\tau = \tau_y$, we have $y_e = y_0 + \Delta x$, and can we solve for x_e from Eq. (2.176) as

$$x_e = x_0 + \frac{1}{g_x}\{[\tilde{v}_{x_0} + g_x(x_i - x_0)]\exp(g_x\Delta\tau) - \tilde{v}_{x_0}\}. \quad (2.181)$$

Similarly, if $\Delta\tau = \tau_x$, we can solve from Eq. (2.177) for y_e as

$$y_e = y_0 + \frac{1}{g_y}\{[\tilde{v}_{y_0} + g_y(y_i - y_0)]\exp(g_y\Delta\tau) - \tilde{v}_{y_0}\}. \quad (2.182)$$

The exit point then forms the entry point of the next grid block and we can repeat the procedure to trace the stream line until it reaches one of the producers. If we sum the travel times over all grid blocks we obtain the *arrival time* for a streamline

Fig. 2.4 Streamlines for
steady state flow in Example 1.
Note: The tracing algorithm
described above computes
parabolic trajectories for the
streamlines in each grid block.
However, the streamlines
have been plotted more
coarsely as straight lines
between the entry and exit
points in the grid blocks

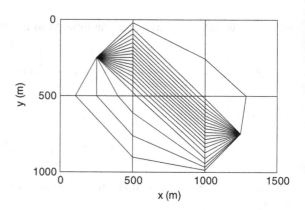

which indicates the moment in time at which a 'virtual' particle travelling with
speed $\tilde{v}_t$ along the streamline would reach the producer (assuming it starts at the
injector at time zero). A related quantity is the *time-of-flight* of a virtual particle
required to reach a specific point along a streamline, which is equal to the summation of grid-block travel times from the injector until that point. Figure 2.4
displays a streamline plot computed for Example 1 after steady state conditions
have been reached. If, for a given total number of streamlines, we choose the
fraction starting from each injector in proportion to the fraction of total water
injected, the distance between streamlines becomes an inverse measure for the
flow per unit surface area (which is also known as the *flux*). In other words, the
closer the streamlines, the higher the flux. Apart from providing a powerful means
to visualize reservoir flow, streamlines can also be used during numerical simulation, as will be briefly discussed in Chap. 3. For a much more in-depth treatment
of streamline methods we refer to the classic papers of Bratvedt, Gimse and
Tegnander (1996), Batycky, Blunt and Thiele (1997), King and Datta Gupta (King
and Datta-Gupta 1998) and to the text book of Datta-Gupta and King (2007).

2.3.7 System Energy*

In analogy to the single-phase case discussed in Sects. 2.2.7 we can formulate the
energy balance per unit time, i.e. the power balance, for the two-phase case. The
power balance can be expressed in terms of potential energy rate, dissipation rate
and work, each related to both oil and water flow. Using matrix–vector notation
this results in Eq. (2.183) below, where we applied the vector and matrix partitioning as introduced in Sects. 2.2.3 to distinguish between gridblocks without
wells, gridblocks with wells where the flow rates are prescribed, and those with
wells were the bottom-hole pressures are prescribed. We note that the presence of
gravity forces and capillary pressures would make the expression for the power
balance more complex. Using a similar reasoning as in Sects. 2.2.7 we can recover

system Eq. (2.135) by first simplifying Eq. (2.183) such that the well index matrices $\mathbf{J}_2$ and $\mathbf{J}_3$ are eliminated, then take derivatives with respect to the state variables $\mathbf{p}_1$, $\mathbf{p}_2$ and $\mathbf{p}_3$, set the result equal to zero, and finally re-introduce the well indices for the prescribed pressures. In Sect. 3.4.5 we will present a numerical example that illustrates the relative importance of the various terms in the power balance.

$$
\begin{aligned}
&[\mathbf{p}_1 \quad \mathbf{p}_2 \quad \mathbf{p}_3] \underbrace{\left\{ \begin{bmatrix} \mathbf{V}_{wp,11} & 0 & 0 \\ 0 & \mathbf{V}_{wp,22} & 0 \\ 0 & 0 & \mathbf{V}_{wp,33} \end{bmatrix} \begin{bmatrix} \dot{\mathbf{p}}_1 \\ \dot{\mathbf{p}}_2 \\ \dot{\mathbf{p}}_3 \end{bmatrix} + \begin{bmatrix} \mathbf{V}_{ws,11} & 0 & 0 \\ 0 & \mathbf{V}_{ws,22} & 0 \\ 0 & 0 & \mathbf{V}_{ws,33} \end{bmatrix} \begin{bmatrix} \dot{\mathbf{s}}_1 \\ \dot{\mathbf{s}}_2 \\ \dot{\mathbf{s}}_3 \end{bmatrix} \right\}}_{\frac{dE_{pot}}{dt}\big|_w} \\
&+[\mathbf{p}_1 \quad \mathbf{p}_2 \quad \mathbf{p}_3] \underbrace{\left\{ \begin{bmatrix} \mathbf{V}_{op,11} & 0 & 0 \\ 0 & \mathbf{V}_{op,22} & 0 \\ 0 & 0 & \mathbf{V}_{op,33} \end{bmatrix} \begin{bmatrix} \dot{\mathbf{p}}_1 \\ \dot{\mathbf{p}}_2 \\ \dot{\mathbf{p}}_3 \end{bmatrix} + \begin{bmatrix} \mathbf{V}_{os,11} & 0 & 0 \\ 0 & \mathbf{V}_{os,22} & 0 \\ 0 & 0 & \mathbf{V}_{os,33} \end{bmatrix} \begin{bmatrix} \dot{\mathbf{s}}_1 \\ \dot{\mathbf{s}}_2 \\ \dot{\mathbf{s}}_3 \end{bmatrix} \right\}}_{\frac{dE_{pot}}{dt}\big|_o} \\
&+\underbrace{[(\mathbf{p}_1 - \mathbf{p}_{av}) \quad (\mathbf{p}_2 - \mathbf{p}_{av}) \quad (\mathbf{p}_3 - \mathbf{p}_{av})] \begin{bmatrix} \mathbf{T}_{w,11} & \mathbf{T}_{w,12} & \mathbf{T}_{w,13} \\ \mathbf{T}_{w,21} & \mathbf{T}_{w,22} & \mathbf{T}_{w,23} \\ \mathbf{T}_{w,31} & \mathbf{T}_{w,32} & \mathbf{T}_{w,33} \end{bmatrix} \begin{bmatrix} \mathbf{p}_1 \\ \mathbf{p}_2 \\ \mathbf{p}_3 \end{bmatrix}}_{\frac{dE_{dis,gb}}{dt}\big|_w} \\
&+\underbrace{[(\mathbf{p}_1 - \mathbf{p}_{av}) \quad (\mathbf{p}_2 - \mathbf{p}_{av}) \quad (\mathbf{p}_3 - \mathbf{p}_{av})] \begin{bmatrix} \mathbf{T}_{o,11} & \mathbf{T}_{o,12} & \mathbf{T}_{o,13} \\ \mathbf{T}_{o,21} & \mathbf{T}_{o,22} & \mathbf{T}_{o,23} \\ \mathbf{T}_{o,31} & \mathbf{T}_{o,32} & \mathbf{T}_{o,33} \end{bmatrix} \begin{bmatrix} \mathbf{p}_1 \\ \mathbf{p}_2 \\ \mathbf{p}_3 \end{bmatrix}}_{\frac{dE_{dis,gb}}{dt}\big|_o} \\
&+\underbrace{\left[0 \quad (\bar{\mathbf{p}}_{well} - \mathbf{p}_2) \quad (\breve{\mathbf{p}}_{well} - \mathbf{p}_3)\right] \begin{bmatrix} 0 & 0 & 0 \\ 0 & \mathbf{F}_{w,22} & 0 \\ 0 & 0 & \mathbf{F}_{w,33} \end{bmatrix} \begin{bmatrix} 0 & 0 & 0 \\ 0 & \mathbf{J}_2 & 0 \\ 0 & 0 & \mathbf{J}_3 \end{bmatrix} \begin{bmatrix} 0 \\ \bar{\mathbf{p}}_{well} - \mathbf{p}_2 \\ \breve{\mathbf{p}}_{well} - \mathbf{p}_3 \end{bmatrix}}_{\frac{dE_{dis,well}}{dt}\big|_w} \\
&+\underbrace{\left[0 \quad (\bar{\mathbf{p}}_{well} - \mathbf{p}_2) \quad (\breve{\mathbf{p}}_{well} - \mathbf{p}_3)\right] \begin{bmatrix} 0 & 0 & 0 \\ 0 & \mathbf{F}_{o,22} & 0 \\ 0 & 0 & \mathbf{F}_{o,33} \end{bmatrix} \begin{bmatrix} 0 & 0 & 0 \\ 0 & \mathbf{J}_2 & 0 \\ 0 & 0 & \mathbf{J}_3 \end{bmatrix} \begin{bmatrix} 0 \\ \bar{\mathbf{p}}_{well} - \mathbf{p}_2 \\ \breve{\mathbf{p}}_{well} - \mathbf{p}_3 \end{bmatrix}}_{\frac{dE_{dis,well}}{dt}\big|_o} \\
&=\underbrace{[0 \quad \bar{\mathbf{p}}_{well} \quad \bar{\mathbf{q}}_{well}] \begin{bmatrix} 0 & 0 & 0 \\ 0 & \mathbf{F}_{w,22} & 0 \\ 0 & 0 & \mathbf{F}_{w,33} \end{bmatrix} \begin{bmatrix} 0 \\ \breve{\mathbf{q}}_{well} \\ \breve{\mathbf{p}}_{well} \end{bmatrix}}_{P_{well}|_w} + \underbrace{[0 \quad \bar{\mathbf{p}}_{well} \quad \bar{\mathbf{q}}_{well}] \begin{bmatrix} 0 & 0 & 0 \\ 0 & \mathbf{F}_{o,22} & 0 \\ 0 & 0 & \mathbf{F}_{o,33} \end{bmatrix} \begin{bmatrix} 0 \\ \breve{\mathbf{q}}_{well} \\ \breve{\mathbf{p}}_{well} \end{bmatrix}}_{P_{well}|_o}.
\end{aligned}
$$

$$(2.183)$$

References

Batycky RP, Blunt MJ, Thiele MR (1997) A 3D field-scale streamline-based reservoir simulator. SPE Reserv Eng 12(4):246–254. doi:10.2118/36726-PA

Bratvedt F, Gimse T, Tegnander C (1996) Streamline computations for porous media flow including gravity. Transp Porous Media 25(1):63–78. doi:10.1007/BF00141262

Brill JP, Mukherjee H (1999) Multi-phase flow in wells. SPE Monograph Series 17, SPE, Richardson

Datta-Gupta A, King MJ (2007) Streamline simulation: Theory and Practice, SPE Textbook Series, 11. SPE, Richardson

Friedland B (1986) Control system design—An introduction to state-space methods. McGraw-Hill, Reprinted in 2005 by Dover, New York

Hubbert MK (1940) The theory of ground-water motion. J Geol 48(8):785–944

Karney BW, Seneviratne A (1991) Application of energy concepts to groundwater flow: time step control and integrated sensitivity analysis. Water Resour Res 27(12):3225–3235

King MJ, Datta-Gupta A (1998) Streamline simulation: a current perspective. In Situ 22(1):91–140

Lanczos C (1970) The variational principles of mechanics, 4th edn. University of Toronto Press, Toronto. Reprinted in 1986 by Dover, New York

Langhaar HL (1962) Energy methods in applied mechanics. Wiley, New York

Oliver DS, Reynolds AC, Liu N (2008) Inverse theory for petroleum reservoir characterization and history matching. Cambridge University Press, Cambridge

Pollock DW (1988) Semi analytical computation of path lines for finite-difference models. Ground Water 26(6):743–750

Zienckiewicz OC, Taylor RL (1989) The finite element method, 4th ed., vol 1–2. McGraw-Hill, London

Chapter 3
System Response

Abstract This chapter first treats the analytical solution of linear systems of ordinary-differential equations for single-phase flow. Next it moves on to the numerical solution of (nonlinear) two-phase flow equations, covering various aspects like implicit, explicit or mixed (IMPES) time discretizations and associated stability issues, Newton-Raphson iteration, streamline simulation, automatic time-stepping, and other computational aspects. The chapter concludes with simple numerical examples to illustrate these and other aspects such as mobility effects, well-constraint switching, time-stepping statistics, and system-energy accounting.

3.1 Free Response

3.1.1 Homogeneous Equation

Consider the linear or linearized time-invariant state-space equations given in Eq. (2.8) or (2.24). The scalar equivalent of these vector differential equations is given by

$$\dot{x}(t) = ax(t) + bu(t) , \qquad (3.1)$$

where a and b are now time-invariant scalar coefficients.[1] Because Eq. (3.1) is first-order in the dependent variable t, it requires a single initial condition:

$$t = \breve{t} : x = \breve{x} . \qquad (3.2)$$

If we set $u = 0$ in Eq. (3.1), we obtain the *homogeneous equation*

$$\dot{x}(t) = ax(t) , \qquad (3.3)$$

---

[1] For an engineering-oriented overview of the theory of first-order scalar and vector ODEs see e.g. the review in Luenberger (1979) or, somewhat more extensively, Boyce and Di Prima (2005). An enormous amount of other textbooks is available covering similar material.

J. D. Jansen, *A Systems Description of Flow Through Porous Media*,
SpringerBriefs in Earth Sciences, DOI: 10.1007/978-3-319-00260-6_3,
© The Author(s) 2013

which describes the *free* response (also called the *transient response*) of the system starting from a non-zero initial condition

$$x(t) = e^{a(t-\breve{t})}\breve{x} \,. \tag{3.4}$$

For values of the coefficient a smaller than zero, the response $x(t)$ for the limit of t approaching infinity becomes zero, i.e. the response is truly transient. For values of a larger than zero, the response grows to infinity, while for $a = 0$, the response remains equal to the initial condition $\breve{x}$. Just as for scalar ODEs, if we set $\mathbf{u} = \mathbf{0}$, in Eq. (3.1) we obtain the homogeneous equation

$$\dot{\mathbf{x}}(t) = \mathbf{A}\mathbf{x}(t) \,, \tag{3.5}$$

with a corresponding initial condition

$$t = \breve{t} : \mathbf{x} = \breve{\mathbf{x}} \,. \tag{3.6}$$

3.1.2 Diagonalization

A solution to Eq. (3.5) can be obtained through *diagonalization* of matrix $\mathbf{A}$. If $\mathbf{A}$ is diagonalizable there exists a non-singular matrix $\mathbf{M}$ of eigenvectors $\mathbf{m}$ of $\mathbf{A}$ such that

$$\mathbf{A} = \mathbf{M}\boldsymbol{\Lambda}\mathbf{M}^{-1} \,, \tag{3.7}$$

where $\boldsymbol{\Lambda} = \text{diag}\,(\lambda_1, \ldots, \lambda_n)$ is the diagonal matrix of eigenvalues of $\mathbf{A}$. Using this decomposition of $\mathbf{A}$, Eq. (3.5) can be written as

$$\dot{\mathbf{x}}(t) = \mathbf{M}\boldsymbol{\Lambda}\mathbf{M}^{-1}\mathbf{x}(t) \,, \tag{3.8}$$

or equivalently, after pre-multiplying (3.8) by $\mathbf{M}^{-1}$, as

$$\mathbf{M}^{-1}\dot{\mathbf{x}}(t) = \boldsymbol{\Lambda}\mathbf{M}^{-1}\mathbf{x}(t) \,. \tag{3.9}$$

Defining a transformed state variable $\mathbf{z}$ as

$$\mathbf{z}(t) = \mathbf{M}^{-1}\mathbf{x}(t) \,, \tag{3.10}$$

and substituting it into (3.9) yields the *decoupled* system of homogeneous equations

$$\dot{\mathbf{z}}(t) = \boldsymbol{\Lambda}\mathbf{z}(t) \,. \tag{3.11}$$

It is called decoupled since each of the elements z_i of the transformed state $\mathbf{z}$ is given by

$$\dot{z}_i(t) = \lambda_i z_i(t) \,, \tag{3.12}$$

thus without being influenced by any of the other elements of $\mathbf{z}$. The solution of (3.12) is given by (c.f. Eq. (3.4))

$$z_i(t) = e^{\lambda_i(t-\breve{t})}\breve{z}_i \,, \tag{3.13}$$

and, equivalently, the solution of the 'full' transformed state variable $\mathbf{z}$ can be written as

$$\mathbf{z}(t) = e^{\Lambda(t-\breve{t})}\breve{\mathbf{z}} \,. \tag{3.14}$$

An interpretation of the matrix exponential e^{Λ} can be obtained by considering the Taylor expansion around zero for the exponential function,

$$e^{\lambda} = 1 + \lambda + \frac{\lambda^2}{2!} + \frac{\lambda^3}{3!} + \dots \,, \tag{3.15}$$

of which the matrix equivalent can be written as[2]

$$e^{\Lambda} = \mathbf{I} + \Lambda + \frac{\Lambda^2}{2!} + \frac{\Lambda^3}{3!} + \dots. \tag{3.16}$$

To recover the solution in terms of the original state variable $\mathbf{x}$, first substitute relationship (3.10) in Eq. (3.14) and multiply with $\mathbf{M}$ to obtain

$$\mathbf{x}(t) = \mathbf{M}e^{\Lambda(t-\breve{t})}\mathbf{M}^{-1}\breve{\mathbf{x}}. \tag{3.17}$$

Using Eqs. (3.7) and (3.16) we can then write

$$\mathbf{M}e^{\Lambda(t-\breve{t})}\mathbf{M}^{-1} = \mathbf{M}\left(\mathbf{I} + \Lambda\left(t-\breve{t}\right) + \frac{\left[\Lambda\left(t-\breve{t}\right)\right]^2}{2!} + \frac{\left[\Lambda\left(t-\breve{t}\right)\right]^3}{3!} + \dots\right)\mathbf{M}^{-1}$$

$$= \mathbf{M}\left(\mathbf{I} + \mathbf{M}^{-1}\mathbf{A}\mathbf{M}\left(t-\breve{t}\right) + \frac{\left[\mathbf{M}^{-1}\mathbf{A}\mathbf{M}\left(t-\breve{t}\right)\right]^2}{2!} + \frac{\left[\mathbf{M}^{-1}\mathbf{A}\mathbf{M}\left(t-\breve{t}\right)\right]^3}{3!} + \dots\right)\mathbf{M}^{-1}$$

$$= \mathbf{I} + \mathbf{A}\left(t-\breve{t}\right) + \frac{\left[\mathbf{A}\left(t-\breve{t}\right)\right]^2}{2!} + \frac{\left[\mathbf{A}\left(t-\breve{t}\right)\right]^3}{3!} + \dots$$

$$= e^{\mathbf{A}\left(t-\breve{t}\right)}, \tag{3.18}$$

and substitution of this relationship in Eq. (3.17) finally gives the solution of homogeneous Eq. (3.5) as

---

[2] In practice, the computation of a matrix exponential should not be performed using this expression; see Moler and Van Loan (1978) for an overview of various possible methods.

$$\mathbf{x}(t) = e^{\mathbf{A}(t-\breve{t})}\breve{\mathbf{x}}. \tag{3.19}$$

Equation (3.19) is the matrix equivalent to scalar Eq. (3.4) and represents the transient behavior of the LTI system with system matrix $\mathbf{A}$. The diagonalization of $\mathbf{A}$ in Eq. (3.7) is an example of a *similarity transformation* because the dynamic system characterized by the transformed system matrix Λ has the same dynamic properties as the one represented by the original system matrix $\mathbf{A}$, since both matrices have the same eigenvalues.

3.1.3 Stability

If all eigenvalues λ_i of matrix $\mathbf{A}$ are smaller than zero, it follows that the response $\mathbf{x}(t)$ of homogeneous Eq. (3.5) for the limit of t approaching infinity becomes zero, i.e. the response is truly transient.[3] This property of a dynamic system is known as *asymptotic stability*. If any of the eigenvalues is larger than zero, the response grows to infinity (that is, in the linear theory), i.e. the system is *unstable*. If at least one of the eigenvalues is equal to zero, whereas the others are smaller than zero, the response for large values of t may approach a non-zero steady-state value, a situation known as *marginal stability*. Physical instability requires an internal source of energy in the system. In the case of flow through porous media such a source normally does not exist, and to the contrary, the system is continuously loosing energy through friction of the fluid in the pores as described by Darcy's law. Unlike in classic control engineering, *physical* instability in time is therefore normally not an issue.[4] An exception is the behavior of coupled well-bore-reservoir systems where occasionally unstable behavior of the well-bore flow may lead to large pressure and flow oscillations in the well-bore and the near-well-bore area. In that case, the source of the instability is in the multiphase flow behavior in the well bore, and not in the reservoir. Analysis of this type of coupled problems requires a

---

[3] More precisely, the condition for asymptotic stability requires that all *real parts of the eigenvalues* are smaller than zero, a condition that is also referred to as the system matrix being Hurwitz. This condition is only of relevance if the eigenvalues are complex numbers, which implies that the system displays oscillatory behavior. However, because we don't take inertia into account in the description of porous-media flow, the system cannot store kinetic energy and the pressures will only display exponentially decaying behavior. Correspondingly, the eigenvalues are real numbers and it suffices to require them to be negative-valued to guarantee asymptotic stability.

[4] However, instabilities in *space* do play a role in reservoir engineering, at least in theory. A well-known case is reservoir flooding with an unfavorable *mobility ratio*, i.e. with the mobility of the displacing fluid being lower than the mobility of the displaced fluid. In that case, a displacement front may become unstable such that *viscous fingering* takes place of the displacing fluid in the displaced fluid. Similar instabilities may occur when a heavy fluid is injected on top of a lighter one, which may lead to fingering caused by *buoyancy-driven convection*. In practice, geological heterogeneities often completely mask the fingering process.

dynamic well-bore simulator that is capable of computing the well-bore dynamics at a time scale of seconds to hours. We will not consider such short-term phenomena, and will restrict our attention to reservoir flow at time scales from days to decades where well-bore flow instabilities play no role. A completely different, artificial, source of instability is related to the numerical simulation of reservoir flow. As will be illustrated in the next section, incorrect time discretization of the system equations may lead to *numerical* instabilities which may completely ruin the simulation, or worse, produce output that at first sight looks in order but contains fluctuations that are unphysical.

3.1.4 Singular System Matrix

In Sect. 2.2.3 it was discussed that the elements of an input vector **u** may consist of prescribed flow rates or prescribed bottom-hole pressures (or a relation between flow rates and pressures). In Sect. 2.2.4 it was shown that the use of a well model to prescribe the bottom-hole pressures results in the addition of a term $\mathbf{J}_p$ to the main block diagonal of the transmissibility matrix **T**, and here we will have a look at an important consequence of that addition. The transmissibility matrix **T** of a reservoir model with only prescribed flow rates, and therefore no prescribed bottom-hole pressures, is singular. This can be understood by considering that **T** defines the transmissibilities between the grid blocks, which directly correspond to the steady-state pressure differences between the grid blocks. However, knowing only the pressure *differences* does not give us enough information to compute the absolute pressures in the grid blocks. This implies that the transmissibility matrix is singular with rank deficiency one. Another way to understand this is by considering the structure of **T**: the sum of every row adds up to exactly zero because of the way the transmissibilities enter the matrix; see Eq. (1.31). Therefore the sum of all columns has to be equal to the zero vector which implies that nontrivial solutions of the homogeneous equation

$$\mathbf{T}\mathbf{p} = \mathbf{0} , \qquad (3.20)$$

are given by

$$\mathbf{p} = p\mathbf{1} , \qquad (3.21)$$

where p is an arbitrary constant pressure. In other words, the null space of **T** consists of all vectors **p** with arbitrary, equal values p in each grid block. To restore regularity of **T** we need to fix at least one of the pressures. Because the well index matrix $\mathbf{J}_3$ is diagonal, addition of $\mathbf{J}_3$ to the main block diagonal of **T** results in the addition of non-zero elements to the main diagonal of **T** which indeed restores the regularity. It will be shown in Sect. 3.2 below that singularity of **T**, and thus of **A**, makes it impossible to directly compute the long-term steady-state pressure distribution in the system, or the behavior in the limit of incompressible flow. It will be shown below that it may still be possible to compute the pressures dynamically through numerical integration of the system equations in time, as long as it concerns compressible flow. However, in that case the singularity of **A** could still lead to numerical problems, in particular for long integration times and small compressibilities.

3.1.5 Example 1 Continued: Free Response

The eigenvalues and eigenvectors of the system matrix **A** can be computed as

$$\Lambda = 10^{-5}$$
$$\times \begin{bmatrix} -0.4163 & 0 & 0 & 0 & 0 & 0 \\ 0 & -0.1723 & 0 & 0 & 0 & 0 \\ 0 & 0 & -0.0660 & 0 & 0 & 0 \\ 0 & 0 & 0 & -0.0285 & 0 & 0 \\ 0 & 0 & 0 & 0 & -0.0059 & 0 \\ 0 & 0 & 0 & 0 & 0 & -0.0000 \end{bmatrix},$$

(3.22)

$$\mathbf{M} = \begin{bmatrix} 0.4108 & 0.6716 & 0.3437 & 0.2351 & 0.2001 & -0.4082 \\ -0.7867 & -0.0760 & 0.3832 & 0.1631 & 0.1875 & -0.4082 \\ 0.0052 & -0.0006 & -0.0107 & 0.0737 & -0.9098 & -0.4082 \\ -0.0571 & 0.0108 & -0.8093 & 0.3551 & 0.2211 & -0.4082 \\ 0.4562 & -0.7269 & 0.2409 & 0.0581 & 0.1881 & -0.4082 \\ -0.0284 & 0.1210 & -0.1479 & -0.8850 & 0.1130 & -0.4082 \end{bmatrix}.$$

(3.23)

Note that one of the eigenvalues (the sixth) is zero, in line with the rank-1 deficiency of **A**. Figure 3.1 displays the eigenvectors by plotting their elements on the corresponding grid blocks. It can be seen that the eigenvector belonging to the zero-eigenvalue has an (arbitrary) constant value.[5] The other five eigenvectors form basis functions in the form of spatial patterns of grid-block pressures.

Next, we perform an analytical integration of the state equations of Example 1 according to Eq. (3.17). Starting from an unbalanced initial condition[6]

---

[5] Eigenvectors are defined up to an arbitrary constant.

[6] I.e. an initial condition where not all grid-block pressures have identical values. The initial condition in Eq. (3.24) consists of spatial fluctuations around the initial reservoir pressure $\bar{p}_R = 30 \times 10^6$ specified in Table 1.1.

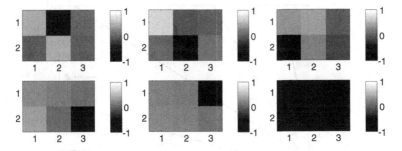

Fig. 3.1 Eigenvectors of the system matrix **A** of Example 1

$$\mathbf{x}(0) = \breve{p}_R \begin{bmatrix} 0.700 \\ 0.900 \\ 1.100 \\ 0.900 \\ 1.100 \\ 1.300 \end{bmatrix} = \begin{bmatrix} 21.000 \\ 27.000 \\ 33.000 \\ 27.000 \\ 33.000 \\ 39.000 \end{bmatrix} \times 10^6 \mathrm{Pa} \ . \tag{3.24}$$

we can compute the response for various values of time t. The initial condition in terms of the coefficient vector $\mathbf{z}$ can be computed as (see Eq. (3.10))

$$\mathbf{Mz}(0) = \mathbf{x}(0) \tag{3.25}$$

which leads to

$$\mathbf{z}(0) = \begin{bmatrix} -0.037 \\ -6.943 \\ -2.455 \\ -11.240 \\ -4.175 \\ -73.485 \end{bmatrix} \times 10^6 \ . \tag{3.26}$$

Note that the coefficients $\mathbf{z}$ are dimensionless which implies that the elements of the eigenvectors $\mathbf{m}$ must have a dimension of pressure (Pa). Figure 3.2 displays the values of the 6 coefficients (i.e. the 6 elements of $\mathbf{z}$) on a logarithmic scale in time, and after one year they have the values

$$\mathbf{z} \underbrace{(365 \times 24 \times 3600)}_{\text{one year in seconds}} = \begin{bmatrix} -0.000 \\ -0.000 \\ -0.000 \\ -0.000 \\ -0.651 \\ -73.485 \end{bmatrix} \times 10^6 \ . \tag{3.27}$$

Clearly the importance of the eigenvectors corresponding to the eigenvalues with the largest absolute values reduces the fastest. The physical interpretation is that

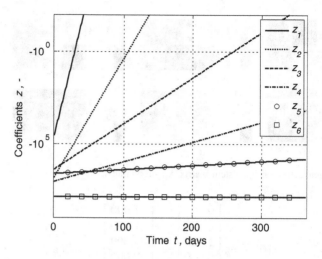

Fig. 3.2 Coefficients z_1 to z_6 as a function of time on a logarithmic scale

those 'modes' are the most heavily damped, and the straight lines in the semi-logarithmic plot illustrate the rapid, exponential nature of the decay. Only the value of the sixth coefficient, multiplying the pattern corresponding to the zero-eigenvalue, does not change its value at all. The product of this coefficient with its corresponding pattern represents the average pressure in all grid blocks after the effect of the initial conditions has been dampened out completely[7]:

$$
\mathbf{x}(\infty) = \mathbf{m}_6 z_6(\infty) =
\begin{bmatrix}
-0.4082 \\
-0.4082 \\
-0.4082 \\
-0.4082 \\
-0.4082 \\
-0.4082
\end{bmatrix}
\times -73.485 \times 10^6 =
\begin{bmatrix}
30.00 \\
30.00 \\
30.00 \\
30.00 \\
30.00 \\
30.00
\end{bmatrix}
\times 10^6 \ \text{Pa} .
$$

(3.28)

The pressure vector $\mathbf{x}$ for the entire period can be recovered as

$$
\mathbf{x}(t) = \mathbf{M}\mathbf{z}(t) ,
$$

(3.29)

and Fig. 3.3 displays the results for the pressures in grid blocks 1, 2, 5 and 6 for a period of one year, which in the end all reach the average value of 30×10^6 Pa.

---

[7] Here we use the notation $\mathbf{x}(\infty)$ to indicate $\lim\limits_{x \to \infty} \mathbf{x}(t)$.

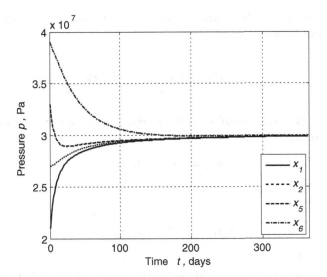

Fig. 3.3 Pressures in grid blocks 1, 2, 5 and 6 as a function of time obtained by analytical integration of the state equations for Example 1

3.2 Forced Response

3.2.1 Nonhomogeneous Equation

We return to the scalar LTI nonhomogeneous Eq. (3.1). Following the standard theory of linear differential equations, we can write the *general solution* of Eq. (3.1) as the sum of the free response, given by Eq. (3.4), and the *forced response* which depends on the *input term* (also known as *forcing term*) $bu(t)$ and which is often referred to as the *particular solution*

$$x(t) = \int_{\underset{\sim}{t}}^{t} e^{a(t-\tau)} bu(\tau) d\tau \,, \tag{3.30}$$

We can interpret the integral in expression (3.30) as the limit for $\Delta\tau \to 0$ of a summation of transient responses to inputs $u(\tau)$, multiplied with b, over small time intervals $\Delta\tau$ during the period $\underset{\sim}{t} \le \tau \le t$. Although mathematically there is no problem in considering cases where time runs backwards, we usually restrict the analysis to cases where the underlying physics forces *causality*, which implies that the states and the outputs are only influenced by past inputs.[8] The general solution of Eq. (3.1) is now obtained as the sum of the homogeneous solution and the

---

[8] A system where the states are influenced by future inputs is therefore referred to as *non-causal*.

particular solution; i.e. the general response is the sum of the transient response and the forced response:

$$x(t) = e^{a(t-\breve{t})}\breve{x} + \int\limits_{\breve{t}}^{t} e^{a(t-\tau)}bu(\tau)d\tau \ . \tag{3.31}$$

In analogy to these scalar results, the forced response of the nonhomogeneous LTI vector differential equation

$$\dot{\mathbf{x}}(t) = \mathbf{A}\mathbf{x}(t) + \mathbf{B}\mathbf{u}(t) \ , \tag{3.32}$$

is given by

$$\mathbf{x}(t) = \int\limits_{\breve{t}}^{t} e^{\mathbf{A}(t-\tau)}\mathbf{B}\mathbf{u}(\tau)d\tau \ , \tag{3.33}$$

such that the general solution is obtained as the sum of solutions (3.19) and (3.33):

$$\mathbf{x}(t) = e^{\mathbf{A}(t-\breve{t})}\breve{\mathbf{x}} + \int\limits_{\breve{t}}^{t} e^{\mathbf{A}(t-\tau)}\mathbf{B}\mathbf{u}(\tau)d\tau \ . \tag{3.34}$$

3.2.2 Diagonalization and Modal Analysis

Just like in the homogeneous case, the inhomogeneous equations may be decoupled through diagonalization of the system equations. Substitution of Eq. (3.7) in Eq. (3.32) results in

$$\dot{\mathbf{x}}(t) = \mathbf{M}\mathbf{\Lambda}\mathbf{M}^{-1}\mathbf{x}(t) + \mathbf{B}\mathbf{u}(t) \ , \tag{3.35}$$

which, after pre-multiplication with $\mathbf{M}^{-1}$ can be written as

$$\dot{\mathbf{z}}(t) = \mathbf{\Lambda}\mathbf{z}(t) + \mathbf{M}^{-1}\mathbf{B}\mathbf{u}(t) \ , \tag{3.36}$$

where $\mathbf{z}$ is a transformed state variable as defined before in Eq. (3.10). The eigenvectors $\mathbf{m}$, i.e. the columns of $\mathbf{M}$, are also known as the *modes* of the dynamic system, and Eq. (3.36) is therefore referred to as a *modal representation* of the system equations. The general solution, Eq. (3.34), can be rewritten in modal form as

$$\mathbf{z}(t) = e^{\mathbf{\Lambda}(t-\breve{t})}\breve{\mathbf{z}} + \int\limits_{\breve{t}}^{t} e^{\mathbf{\Lambda}(t-\tau)}\mathbf{M}^{-1}\mathbf{B}\mathbf{u}(\tau)d\tau \ . \tag{3.37}$$

In case of physical systems where inertia plays a role, such as e.g. mechanical (mass-spring) systems or electrical (inductance-capacitance) networks, the modes correspond to spatial patterns of oscillations for the undamped homogeneous system. In particular for mechanical systems there exists an extensive branch of *modal analysis* techniques to obtain the modes (eigenvectors) and the associated frequencies (eigenvalues) of a system from experiments. In the case of flow through porous media, inertia is usually neglected, which means that the free response of the system is non-oscillatory and just consists of decaying exponential functions (see Eq. (3.19)), such that the modes have much less physical significance. Note that although the homogeneous equations were fully decoupled (see Eqs. (3.11) and (3.12)), the inhomogeneous equations are in general coupled through the input, because except for the special case that $\mathbf{M}^{-1}\mathbf{B}$ is a unit matrix, the elements of the input vector $\mathbf{u}$ will influence more than just a single mode.

3.2.3 Singular System Matrix

3.2.3.1 Steady-State Response

Consider the LTI inhomogeneous Eq. (3.32) with a regular system matrix $\mathbf{A}$. In the special case that the input vector $\mathbf{u}(t)$ becomes a constant $\mathbf{u}(\infty)$ for $t \to \infty$, the steady-state solution can be obtained by putting $\dot{\mathbf{x}} = \mathbf{0}$, resulting in the linear system of equations

$$\mathbf{A}\mathbf{x}(\infty) = -\mathbf{B}\mathbf{u}(\infty) , \tag{3.38}$$

which could then be solved for $\mathbf{x}(\infty)$. Formally this can be written as

$$\mathbf{x}(\infty) = -\mathbf{A}^{-1}\mathbf{B}\mathbf{u}(\infty) , \tag{3.39}$$

although in practice it is computationally more efficient to solve the system of Eqs. (3.38). Note, however, that for a singular matrix $\mathbf{A}$ we cannot solve Eq. (3.38). As discussed in Sect. 3.1.4 above, $\mathbf{T}$, and therefore $\mathbf{A}$, are singular if we prescribe only the flow rates in the wells. If we fix at least one of the pressures with the aid of a well model the resulting modified transmissibility matrix $(\mathbf{T}^* + \mathbf{J}_p^*)$ is regular, and therefore also the system matrix $\mathbf{A}^*$, and we can compute the steady-state vectors $\mathbf{x}^*(\infty)$ and $\mathbf{y}^*(\infty)$ from

$$\mathbf{A}^*\mathbf{x}^*(\infty) = -\mathbf{B}^*\mathbf{u}^*(\infty) , \tag{3.40}$$

$$\mathbf{y}^*(\infty) = \mathbf{C}^*\mathbf{x}^*(\infty) + \mathbf{D}^*\mathbf{u}^*(\infty) . \tag{3.41}$$

3.2.3.2 Incompressible Flow

A similar situation occurs in the limit of incompressible flow. As discussed in Sect. 1.3, in that case the accumulation matrix $\mathbf{V}$ vanishes, such that differential Eq. (2.8) is replaced by an algebraic equation

$$\mathbf{A}\mathbf{x}(t) + \mathbf{B}\mathbf{u}(t) = \mathbf{0} , \tag{3.42}$$

with

$$\mathbf{A} = -\mathbf{T} , \tag{3.43}$$

$$\mathbf{B} = \mathbf{L}_{qu} . \tag{3.44}$$

The solution can be obtained by solving the system of equations

$$\mathbf{A}\mathbf{x}(t) = -\mathbf{B}\mathbf{u}(t) , \tag{3.45}$$

where the dynamic response $\mathbf{x}(t)$ is now assumed to occur instantaneously. Just as in the steady-state case, $\mathbf{A}$ needs to be regular, i.e. we need to fix at least one of the pressures in the wells.

3.3 Numerical Simulation

Until now we have considered the response of simple linear reservoir systems for which it was possible to obtain analytical solutions. For more realistic, nonlinear, reservoir systems we need to simulate the response numerically.

3.3.1 Explicit Euler Discretization

To numerically simulate the dynamic system behavior it is required to discretize not only the spatial variables but also the time variable. In other words, we need to discretize the *continuous-time* system of ODEs and derive a *discrete-time* system of ODEs. Starting from the general form of the system equations (2.6), the most simple approach is to discretize the equation by replacing the difference $d\mathbf{x}/dt$ by differential $\Delta\mathbf{x}/\Delta t$:

$$\frac{\Delta\mathbf{x}}{\Delta t} \approx \mathbf{f}(\mathbf{u}(t), \mathbf{x}(t)) . \tag{3.46}$$

This gives us an algorithm to compute an approximate new value $\mathbf{x}_k$ at t_k from a known value $\mathbf{x}_{k-1}$ at t_{k-1}[9]:

---

[9] We use the shortcut notation $\mathbf{x}_k$ to indicate $\mathbf{x}(t_k)$, i.e. the value of $\mathbf{x}$ at $t = t_k$.

$$\mathbf{x}_k = \mathbf{x}_{k-1} + \Delta\mathbf{x} \approx \mathbf{x}_{k-1} + \mathbf{f}(\mathbf{u}_{k-1}, \mathbf{x}_{k-1})\Delta t \,, \qquad (3.47)$$

where $\Delta t = t_k - t_{k-1}$. The counter k is generally referred to as the discrete time. More formally, the same result is obtained by using a forward Taylor expansion for $\mathbf{x}$ at t_{k-1}:

$$\mathbf{x}_k = \mathbf{x}_{k-1} + \left[\frac{d\mathbf{x}(t)}{dt}\right]_{t=t_{k-1}} \Delta t + \frac{1}{2}\left[\frac{d^2\mathbf{x}(t)}{dt^2}\right]_{t=t_{k-1}} (\Delta t)^2 + \dots . \qquad (3.48)$$

Substitution of Eq. (2.5) into Eq. (3.48) and disregarding all terms higher than first-order leads indeed to Eq. (3.47). As an illustration we apply the simple approach to the LTI state and output Eqs. (2.8) and (2.11),

$$\dot{\mathbf{x}}(t) = \mathbf{A}_c\mathbf{x}(t) + \mathbf{B}_c\mathbf{u}(t) \,, \qquad (3.49)$$

$$\mathbf{y}(t) = \mathbf{C}_c\mathbf{x}(t) + \mathbf{D}_c\mathbf{u}(t) \,, \qquad (3.50)$$

where we have now added subscripts c to matrices $\mathbf{A}$, $\mathbf{B}$, $\mathbf{C}$ and $\mathbf{D}$ to indicate that they are related to a continuous-time representation. Following Eq. (3.47) we obtain

$$\mathbf{x}_k = \mathbf{x}_{k-1} + (\mathbf{A}_c\mathbf{x}_{k-1} + \mathbf{B}_c\mathbf{u}_{k-1})\Delta t \,. \qquad (3.51)$$

Defining

$$\mathbf{A}_d = (\mathbf{I} + \Delta t\,\mathbf{A}_c) \,, \qquad (3.52)$$

$$\mathbf{B}_d = \Delta t\,\mathbf{B}_c \,, \qquad (3.53)$$

$$\mathbf{C}_d = \mathbf{C}_c \qquad (3.54)$$

and

$$\mathbf{D}_d = \mathbf{D}_c \,, \qquad (3.55)$$

allows us to write the general state-space system in discrete-time form:

$$\mathbf{x}_k = \mathbf{A}_d\mathbf{x}_{k-1} + \mathbf{B}_d\mathbf{u}_{k-1} \,, \qquad (3.56)$$

$$\mathbf{y}_k = \mathbf{C}_d\mathbf{x}_k + \mathbf{D}_d\mathbf{u}_k \,. \qquad (3.57)$$

Equation (3.56) is a *difference equation* or *discrete-time differential equation*. It represents one particular discrete-time equivalent of the continuous-time differential Eq. (2.8). Many other time discretizations are possible and we will discuss some of them later on in this chapter. Similarly, discrete-time equivalents can be obtained for the LTV and nonlinear continuous-time state equations. For the general case of nonlinear systems with time-varying parameters, the discrete-time equivalent to Eqs. (2.6) and (2.10) can be expressed as

$$\mathbf{x}_k = \mathbf{f}_k(\mathbf{u}_{k-1}, \mathbf{x}_{k-1}) \,, \qquad (3.58)$$

$$\mathbf{y}_k = \mathbf{h}_k(\mathbf{u}_k, \mathbf{x}_k) \ . \tag{3.59}$$

Comparison with Eq. (3.47) shows that this implies that

$$\mathbf{f}_k = \mathbf{x}_{k-1} + \mathbf{f}_{c,k-1}\Delta t \tag{3.60}$$

and

$$\mathbf{h}_k = \mathbf{h}_{c,k} \ , \tag{3.61}$$

where we used subscripts c, $k-1$ and c, k to indicate continuous-time functions evaluated at discrete times $k-1$ and k respectively.

3.3.2 Implicit Euler Discretization

Equation (3.47) is known as an *explicit Euler scheme*, where the term explicit refers to the fact that $\mathbf{x}_{k+1}$ can be obtained as an explicit formula in terms of $\mathbf{x}_k$. This is possible because we chose to evaluate the function $\mathbf{f}$ in Eq. (3.46) at the 'old' time t_{k-1}. If, alternatively, we choose to evaluate $\mathbf{f}$ at the 'new' time t_k and apply the result to the LTI Eq. (2.8) we obtain

$$\mathbf{x}_k = \mathbf{x}_{k-1} + (\mathbf{A}_c\mathbf{x}_k + \mathbf{B}_c\mathbf{u}_k)\Delta t \ , \tag{3.62}$$

As before, a more formal derivation of this result can be obtained by using a Taylor expansion; this time a backward one for $\mathbf{x}$ at t_k:

$$\mathbf{x}_{k-1} = \mathbf{x}_k - \left[\frac{d\mathbf{x}(t)}{dt}\right]_{t=t_k}\Delta t - \frac{1}{2}\left[\frac{d^2\mathbf{x}(t)}{dt^2}\right]_{t=t_k}(\Delta t)^2 - \dots \tag{3.63}$$

Re-ordering the terms leads to:

$$\mathbf{x}_k = \mathbf{x}_{k-1} + \left[\frac{d\mathbf{x}(t)}{dt}\right]_{t=t_k}\Delta t + \frac{1}{2}\left[\frac{d^2\mathbf{x}(t)}{dt^2}\right]_{t=t_k}(\Delta t)^2 + \dots , \tag{3.64}$$

which is identical to expression (3.48) except for the time at which the derivatives are evaluated. Substitution of Eq. (2.8) in Eq. (3.64), and disregarding the terms higher than first order, leads to Eq. (3.62) again. This equation is known as an *implicit Euler scheme*, because $\mathbf{x}_k$ appears both at the left-hand and the right-hand side of the equation. It can be rewritten as

$$(\mathbf{I} - \Delta t\mathbf{A}_c)\mathbf{x}_k = \mathbf{x}_{k-1} + \Delta t\mathbf{B}_c\mathbf{u}_k \ , \tag{3.65}$$

whereafter it can formally be solved for $\mathbf{x}_k$ as

$$\mathbf{x}_k = (\mathbf{I} - \Delta t\mathbf{A}_c)^{-1}(\mathbf{x}_{k-1} + \Delta t\mathbf{B}_c\mathbf{u}_k) \ , \tag{3.66}$$

although in a numerical implementation it is, as always, more efficient to solve the linear system of Eq. (3.65) than to compute the inverse as in Eq. (3.66). Expression (3.66) can be rewritten in a form similar to Eq. (3.56) if we redefine $\mathbf{A}_d$ and $\mathbf{B}_d$ as

$$\mathbf{A}_d = (\mathbf{I} - \Delta t \mathbf{A}_c)^{-1} , \tag{3.67}$$

$$\mathbf{B}_d = (\mathbf{I} - \Delta t \mathbf{A}_c)^{-1} \Delta t \mathbf{B}_c , \tag{3.68}$$

leading to the discrete state-space form

$$\mathbf{x}_k = \mathbf{A}_d \mathbf{x}_{k-1} + \mathbf{B}_d \mathbf{u}_k \tag{3.69}$$

Note that although Eq. (3.69) appears to be explicit in time again, the underlying implicit discretization scheme results in the need to solve a system of equations at each time step. An implicit Euler discretization of the nonlinear system of Eq. (2.6) can also be obtained by substitution in Eq. (3.64). Disregarding higher-order terms, this leads to

$$\mathbf{x}_k = \mathbf{x}_{k-1} + \Delta t \, \mathbf{f}_c(\mathbf{u}_k, \mathbf{x}_k) , \tag{3.70}$$

or, with $\mathbf{f}_k = \mathbf{x}_{k-1} + \mathbf{f}_{c,k} \Delta t$, to

$$\mathbf{x}_k = \mathbf{f}_k(\mathbf{u}_k, \mathbf{x}_{k-1}, \mathbf{x}_k) . \tag{3.71}$$

3.3.3 Picard and Newton–Raphson Iteration

It is, in general, not possible to invert the nonlinear function $\mathbf{f}_k$ and therefore we need an iterative procedure to solve Eq. (3.71) at every time step. The most simple procedure is *Picard iteration*, also known as *subsequent substitution* or *simple iteration*, in which we start by solving the equation with an initial guess $\mathbf{x}_k^0$ at the right-hand-side to obtain an improved estimate $\mathbf{x}_k^1$ at the left-hand side. The usual choice for $\mathbf{x}_k^0$ is simply the value $\mathbf{x}_{k-1}$ computed during the previous time step. Subsequent iteration steps can then be expressed as

$$\mathbf{x}_k^i = \mathbf{f}_k(\mathbf{u}_k, \mathbf{x}_{k-1}, \mathbf{x}_k^{i-1}) , \tag{3.72}$$

where the superscript i is the iteration counter. The iteration is terminated when a predefined convergence criterion is met. A typical criterion is given in terms of the two-norm

$$\left\| \mathbf{r}_k^i \right\|_2 \triangleq \sqrt{\sum_{i=1}^{n} r_i^2} \leq \varepsilon , \tag{3.73}$$

where $\mathbf{r}_k^i = \mathbf{x}_k^i - \mathbf{x}_k^{i-1}$ is the *residual* of the iteration, n is the number of elements in $\mathbf{r}$, and ε is a small number.[10] Another popular norm to specify convergence criteria is the infinity norm

$$\left\| \mathbf{r}_k^i \right\|_\infty \triangleq \max_i (r_i) , \quad i = 1, \ldots, n . \tag{3.74}$$

Expressions (3.73) and (3.74) are known as an *absolute* convergence criteria. An example of a *relative* criterion is

$$\frac{\left\| \mathbf{r}_k^i \right\|_2}{\left\| \mathbf{x}_k^i \right\|_2} \leq \varepsilon . \tag{3.75}$$

In practice, it is often required that several convergence criteria are met simultaneously before an iteration may be terminated. An alternative to Picard iteration is *Newton-Raphson iteration*. The vectorial form of this iteration scheme can be expressed as the two-step procedure:

$$\frac{\partial \mathbf{g}_k \left(\mathbf{u}_k, \mathbf{x}_{k-1}, \mathbf{x}_k^i \right)}{\partial \mathbf{x}_k^i} \mathbf{r}_k^i = \mathbf{g}_k \left(\mathbf{u}_k, \mathbf{x}_{k-1}, \mathbf{x}_k^i \right) , \tag{3.76}$$

$$\mathbf{x}_k^{i+1} = \mathbf{x}_k^i + \mathbf{r}_k^i , \tag{3.77}$$

where

$$\mathbf{g}_k (\mathbf{u}_k, \mathbf{x}_{k-1}, \mathbf{x}_k) \triangleq \mathbf{x}_k - \mathbf{f}_k (\mathbf{u}_k, \mathbf{x}_{k-1}, \mathbf{x}_k) = \mathbf{0} \tag{3.78}$$

is the implicit form of the system Eq. (3.71), and where

$$\frac{\partial \mathbf{g}_k \left(\mathbf{u}_k, \mathbf{x}_{k-1}, \mathbf{x}_k^i \right)}{\partial \mathbf{x}_k^i} = \begin{bmatrix} \frac{\partial g_1}{\partial x_1} & \frac{\partial g_1}{\partial x_2} & \cdots & \frac{\partial g_1}{\partial x_n} \\ \frac{\partial g_2}{\partial x_1} & \frac{\partial g_2}{\partial x_2} & \cdots & \frac{\partial g_2}{\partial x_n} \\ \vdots & \vdots & \ddots & \vdots \\ \frac{\partial g_n}{\partial x_1} & \frac{\partial g_n}{\partial x_2} & \cdots & \frac{\partial g_n}{\partial x_n} \end{bmatrix}_k^i , \tag{3.79}$$

is known as a *Jacobian* matrix. Equation (3.76) implies that we have to solve a system of linear equations to find $\mathbf{r}_k^i$ during each iteration step. Just as for Picard iteration, the convergence criterion for Newton-Raphson iteration can be specified in terms of the residual $\mathbf{r}_k^i$ in various ways. Newton-Raphson iteration generally

---

[10]   The small number ε itself is often also called the convergence criterion. The state vector $\mathbf{x}$ may contain groups of elements with different physical dimensions (e.g. pressures and saturations), in which case the dimensions of ε are ill-defined. Moreover, the magnitudes of the pressures are usually much larger than those of the saturations (typically 10^6–10^7 versus 0–1) and therefore the pressure values determine whether or not the convergence criterion is met whereas the saturations have almost no influence. For multi-phase flow it is therefore required to specify separate convergence criteria for the pressures and the saturations, or to scale the variables such that they become dimensionless and of the same order of magnitude.

converges faster than Picard iteration, especially in the close neighborhood of the root to which it converges. However, both methods may occasionally fail to converge in a reasonable number of iteration steps. Various ad-hoc measures to guide the iteration process, or to restart the process after failure, are therefore usually applied in numerical implementations.

For a reservoir-specific example consider the continuous-time state-space representation for two-phase flow with or without well model as given in Eqs. (2.142) and (2.144):

$$\dot{\mathbf{x}} = \mathbf{A}_c(\mathbf{x})\mathbf{x} + \mathbf{B}_c(\mathbf{x})\mathbf{u} . \tag{3.80}$$

Here we have, as before, added subscripts c to indicate that the secant matrices $\mathbf{A}_c$ and $\mathbf{B}_c$ represent a continuous-time formulation. Applying implicit Euler discretization results in

$$\mathbf{x}_k = \mathbf{x}_{k-1} + \Delta t \mathbf{A}_c(\mathbf{x}_k)\mathbf{x}_k + \Delta t \mathbf{B}_c(\mathbf{x}_k)\mathbf{u}_k , \tag{3.81}$$

or, formally,

$$\mathbf{x}_k = \mathbf{A}_d(\mathbf{x}_k)\mathbf{x}_{k-1} + \mathbf{B}_d(\mathbf{x}_k)\mathbf{u}_k , \tag{3.82}$$

where

$$\mathbf{A}_d(\mathbf{x}_k) = [\mathbf{I} - \Delta t \mathbf{A}_c(\mathbf{x}_k)]^{-1} , \quad \mathbf{B}_d(\mathbf{x}_k) = \Delta t [\mathbf{I} - \Delta t \mathbf{A}_c(\mathbf{x}_k)]^{-1} \mathbf{B}_c(\mathbf{x}_k) . \tag{3.83}$$

If we want to solve Eq. (3.82) using Newton-Raphson iteration, we could, formally, specify the implicit version of Eq. (3.82) in the form of a function $\mathbf{g}_k$ as

$$\mathbf{g}_k(\mathbf{u}_k, \mathbf{x}_{k-1}, \mathbf{x}_k) = \mathbf{x}_k - \mathbf{A}_d(\mathbf{x}_k)\mathbf{x}_{k-1} - \mathbf{B}_d(\mathbf{x}_k)\mathbf{u}_k , \tag{3.84}$$

and work out the Jacobian $\partial \mathbf{g}_k / \partial \mathbf{x}_k$. In practice, it will be more convenient to start from the version with continuous-time matrices, as given in Eq. (3.81), such that $\mathbf{g}_k$ is expressed as:

$$\mathbf{g}_k(\mathbf{u}_k, \mathbf{x}_{k-1}, \mathbf{x}_k) = (\mathbf{I} - \Delta t \mathbf{A}_c(\mathbf{x}_k))\mathbf{x}_k - \mathbf{x}_{k-1} - \Delta t \mathbf{B}_c(\mathbf{x}_k)\mathbf{u}_k . \tag{3.85}$$

Moreover, it is usually computationally more efficient to use the generalized state-space formulation, which leads to

$$\mathbf{g}_k(\mathbf{u}_k, \mathbf{x}_{k-1}, \mathbf{x}_k) = \left(\hat{\mathbf{E}}_c(\mathbf{x}_k) - \Delta t \hat{\mathbf{A}}_c(\mathbf{x}_k) \right) \mathbf{x}_k - \hat{\mathbf{E}}_c(\mathbf{x}_k)\mathbf{x}_{k-1} - \Delta t \hat{\mathbf{B}}_c(\mathbf{x}_k)\mathbf{u}_k . \tag{3.86}$$

3.3.4 Numerical Stability

In Sect. 3.1.3 we addressed the stability of a continuous-time dynamical system, and we found that asymptotic stability requires that all eigenvalues of the system matrix $\mathbf{A}$ are smaller than zero. We also discussed, in Sect. 3.1.2, how a coupled system of n equations $\dot{\mathbf{x}}(t) = \mathbf{A}\mathbf{x}(t)$ can be transformed to a system of n uncoupled equations $\dot{\mathbf{z}}(t) = \Lambda\mathbf{z}(t)$, where $\mathbf{z}(t) = \mathbf{M}^{-1}\mathbf{x}(t)$ and where $\Lambda = \text{diag}(\lambda_1, ..., \lambda_n)$ is the diagonal matrix of eigenvalues of $\mathbf{A}$ with $\lambda_1 \le \lambda_2 \le ... \le \lambda_n$. The stability of the system is therefore governed by the stability of the individual differential equations of the uncoupled system:

$$\dot{z}_i(t) = \lambda_i z_i(t) . \tag{3.87}$$

In particular, if $\lambda_n < 0$, the system is asymptotically stable. If $\lambda_n = 0$, the system is marginally stable. To establish the stability properties of a time-discretized system of equations we will therefore consider the discretized version of Eq. (3.87). E.g. for the explicit Euler case, we can write:

$$z_{i,k} = (1 + \Delta t \lambda_i)z_{i,k-1} . \tag{3.88}$$

This recursive equation will asymptotically approach zero for large values of k if

$$-1 < (1 + \Delta t \lambda_i) < 1 . \tag{3.89}$$

This is equivalent to

$$0 < \Delta t < \frac{2}{-\lambda_i} . \tag{3.90}$$

If we restrict our attention to integration forward in time, i.e. to $\Delta t > 0$, inequalities (3.90) imply that the discretized system is only stable if

1. the underlying continuous-time system is stable, i.e. if $\lambda_i < 0$ for all i, otherwise Δt cannot fulfill both inequalities, and
2. the time step Δt is in between the bounds given by inequalities (3.90). This requirement is known as the *Courant-Friedrichs-Lewy (CFL) stability condition*, and the explicitly discretized system is therefore *conditionally* stable.[11]

In case of explicit Euler discretization the stability is therefore governed by the negative eigenvalue with the largest absolute value, λ_1. In case of implicit Euler discretization we have

$$z_{i,k} = (1 - \Delta t \lambda_i)^{-1}z_{i,k-1} , \tag{3.91}$$

which will be asymptotically stable if

$$1 < |1 - \Delta t \lambda_i| . \tag{3.92}$$

If the underlying system is asymptotically stable, and if we take Δt positive, we find that this condition is always fulfilled. In other words, the implicitly discretized system is *unconditionally stable*. For nonlinear systems we may apply these

---

[11] The upper bound $\Delta t < -2/\lambda_1$ is also known as the *CFL limit*.

concepts to the linearized Eq. (2.24), i.e. to the tangent-linear system, although the results are then only valid for a local neighborhood around each point along the state trajectory. The combined linearized pressure and saturation equations for porous-media flow form a stiff system of equations, i.e. the ratio between the largest and smallest eigenvalues is very big. In particular, the absolute values of the eigenvalues corresponding to the pressure equations are much larger than those corresponding to the saturation equations. As a consequence the time step for explicit integration of the pressure equations becomes so small that for all practical purposes this is not an option. Through analysis of the linearized time-discretized saturation equations it can be shown that the stability limit for explicit integration is governed by grid blocks with the highest *throughput*, i.e. total flow rate per time step; see e.g. Aziz and Settari (1979) or Datta-Gupta and King (2007). In particular, for the case of one-dimensional incompressible two-phase flow without capillary forces the maximum time step for explicit Euler integration of the corresponding Buckley-Leverett equation is governed by the throughput condition:

$$\Delta t < \frac{\Delta x \, \phi (1 - S_{or} - S_{wc})}{v_t v_D^*} , \qquad (3.93)$$

where v_D^* is the dimensionless shock front velocity as defined in Eq. (1.91). Similar expressions can be obtained for more complicated cases. For all cases, higher total fluid velocities and smaller spatial grid-block dimensions imply a smaller time step to maintain stability.

3.3.5 IMPES

Although the implicit formulation allows arbitrarily large time steps as far as stability is concerned, there is usually a time step restriction based on accuracy requirements. In particular, it is often required to restrict the saturation changes per time step such that they stay considerably below one. In that case the time step size is typically below the stability limit for the saturation equations, but above the limit for the pressure equations. A popular alternative for the time integration of equations for multiphase flow through porous media is therefore the *IMplicit Pressure– Explicit Saturation* (IMPES) scheme. In the IMPES scheme the equations are reorganized such that it is possible to solve for pressures and saturations separately, which allows for an implicit update of the pressures, and a stable explicit update of the saturations, using the same time step size. Since explicit updates do not involve the solving of equations, they are much faster than implicit updates, and therefore the IMPES scheme is computationally attractive. Alternatively, it is possible to use a large time step for the implicit pressure update (with a size above the stability limit for the explicit saturation updates), and use a smaller step size to perform multiple explicit saturation updates in-between the pressure updates. To obtain the IMPES formulation, consider again the general continuous-time state-space representation for two-phase flow with or without well model

$$\dot{\mathbf{x}} = \mathbf{A}_c(\mathbf{x})\mathbf{x} + \mathbf{B}_c(\mathbf{x})\mathbf{u} . \qquad (3.94)$$

Recalling that the state vector $\mathbf{x}$ consists of the pressures $\mathbf{p}$ and the saturations $\mathbf{s}$, we may partition Eq. (3.94) as

$$\begin{bmatrix} \dot{\mathbf{p}} \\ \dot{\mathbf{s}} \end{bmatrix} = \begin{bmatrix} \mathbf{A}_p(\mathbf{s}) & 0 \\ \mathbf{A}_s(\mathbf{s}) & 0 \end{bmatrix} \begin{bmatrix} \mathbf{p} \\ \mathbf{s} \end{bmatrix} + \begin{bmatrix} \mathbf{B}_p(\mathbf{s}) \\ \mathbf{B}_s(\mathbf{s}) \end{bmatrix} \mathbf{u} , \qquad (3.95)$$

which can also be written as two separate systems of equations:

$$\dot{\mathbf{p}} = \mathbf{A}_p(\mathbf{s})\mathbf{p} + \mathbf{B}_p(\mathbf{s})\mathbf{u} , \qquad (3.96)$$

$$\dot{\mathbf{s}} = \mathbf{A}_s(\mathbf{s})\mathbf{p} + \mathbf{B}_s(\mathbf{s})\mathbf{u} . \qquad (3.97)$$

Equation (3.96) is a linear differential equation for $\mathbf{p}$ that can be solved implicitly as[12]:

$$\mathbf{p}_k = \left[\mathbf{I} - \Delta t \mathbf{A}_p(\mathbf{s}_{k-1}) \right]^{-1} \left[\mathbf{p}_{k-1} + \Delta t \mathbf{B}_p(\mathbf{s}_{k-1})\mathbf{u}_k \right] \qquad (3.98)$$

or, computationally more efficiently, as

$$\left[\mathbf{I} - \Delta t \mathbf{A}_p(\mathbf{s}_{k-1}) \right] \mathbf{p}_k = \mathbf{p}_{k-1} + \Delta t \mathbf{B}_p(\mathbf{s}_{k-1})\mathbf{u}_k . \qquad (3.99)$$

Equation (3.97) is a nonlinear equation for $\mathbf{s}$, which can be solved explicitly as

$$\mathbf{s}_k = \mathbf{s}_{k-1} + \Delta t \mathbf{A}_s(\mathbf{s}_{k-1})\mathbf{p}_k + \Delta t \mathbf{B}_s(\mathbf{s}_{k-1})\mathbf{u}_k \qquad (3.100)$$

Note that it is possible to use the input at time k in Eqs. (3.99) and (3.100) and the pressure at time k in Eq. (3.100), but that we have to use the saturation at time $k-1$ in Eq. (3.99). Of course it is possible to repeat the implicit pressure computation (3.99) with the new saturation vector $\mathbf{s}_k$ as obtained from Eq. (3.100) and repeat this process until convergence. Moreover, also the saturation update may be performed implicitly, in which case we obtain a scheme known as the *sequential solution method*. For an alternative, more traditional, derivation of the IMPES scheme, and for a detailed analysis of properties such as mass conservation, stability, and accuracy, consult Aziz and Settari (1979) or Chen et al. (2006). In the special case that the fluid and rock compressibilities can be taken as zero, i.e. in the case of *incompressible flow*, the IMPES Eq. (3.99) for the pressures reduces to an algebraic equation

$$\tilde{\mathbf{A}}_p(\mathbf{s}_{k-1})\mathbf{p}_k = \tilde{\mathbf{B}}_p(\mathbf{s}_{k-1})\mathbf{u}_k , \qquad (3.101)$$

where

$$\tilde{\mathbf{A}}_p(\mathbf{s}_{k-1}) = \mathbf{T}_w(\mathbf{s}_{k-1}) + \mathbf{T}_o(\mathbf{s}_{k-1}) , \qquad (3.102)$$

---

[12] In the more general case that $\mathbf{A}_p$ is a continuous function of $\mathbf{p}$, an iterative implicit solution using Picard or Newton iteration will be required.

$$\tilde{\mathbf{B}}_p(\mathbf{s}_{k-1}) = [\mathbf{F}_w(\mathbf{s}_{k-1}) + \mathbf{F}_o(\mathbf{s}_{k-1})]\mathbf{L}_{qu} , \qquad (3.103)$$

as has been derived using the material from Sect. 1.4.11. The incompressible IMPES equation for saturations still has the form of Eq. (3.100), but with modified system and input matrices[13]:

$$\mathbf{A}_s(\mathbf{s}_{k-1}) = \mathbf{V}_{ws}^{-1}\mathbf{T}_w(\mathbf{s}_{k-1}) , \qquad (3.104)$$

$$\mathbf{B}_s(\mathbf{s}_{k-1}) = \mathbf{V}_{ws}^{-1}\mathbf{F}_w(\mathbf{s}_{k-1})\mathbf{L}_{qu} . \qquad (3.105)$$

3.3.6 Computational Aspects

- In most simulators the wells are represented using a well model, and the user can specify either the bottom-hole pressure (BHP) or the total flow rate for each well. In addition it is possible to prescribe a maximum BHP for an injector with a prescribed rate, or a minimum BHP for a producer with a prescribed rate. Similarly, the user can prescribe maximum or minimum rates for wells with prescribed BHPs. Every time step the integration algorithm checks for violation of the constraints, and if so, recomputes the time step with a new prescribed well condition. This implies that a well that is controlled on BHP may become a rate-controlled well or vice versa. In fact the prescription of a rate or BHP can be considered a constraint itself, and the program simply checks that at any moment in time the *most constraining constraint* is active.
- For explicit integration the time step size is limited to maintain stability, as was discussed in Sect. 3.3.4. For implicit integration, which is mostly used in reservoir simulation, the time step limitation is governed by accuracy requirements for which there exist no hard rules. Most simulators employ a variable step size algorithm. An example of such an algorithm is given in Aziz and Settari (1979). It aims at maintaining pressure and saturation changes at or below prescribed levels Δp_{target} and $\Delta S_{w,target}$ for each time step by adjusting the new time step based on the converged results from the previous time step according to:

$$\Delta t_{new} = \min\left[\Delta t_{old}\left(\frac{\Delta p_{target}}{\Delta p}\right), \Delta t_{old}\left(\frac{\Delta S_{w,target}}{\Delta S_w}\right)\right] . \qquad (3.106)$$

Because this involves extrapolation from the previous time step, the actual pressure and saturation changes will sometimes somewhat overshoot the target values. Therefore optional maximum allowed changes may be specified which, if exceeded, will trigger repetition of the integration step with a reduced step

---

[13] Here we have chosen, arbitrarily, to base the incompressible IMPES saturation equation on Eq. (1.132) which is expressed in terms of $\mathbf{V}_{ws}$, $\mathbf{T}_w$ and $\mathbf{F}_w$. We could just as well have used equation (1.133), which would have resulted in an expression in terms of $\mathbf{V}_{os}$, $\mathbf{T}_o$ and $\mathbf{F}_o$.

size. Moreover, the step size may be limited to stay below a maximum allowed
value.

- Solution of the linear system of equations within each Newton-Raphson iteration can be performed with a *direct solution method*, or, for large systems, with
  an *iterative solution method* in conjunction with a *pre-conditioner*. A treatment
  of these numerical mathematics aspects is outside the scope of this text, and we
  refer to e.g. Chen et al. (2006) for a detailed discussion.
- The implicit simulation using Newton-Raphson iterations can often be accel-
  erated by restricting the update of the Jacobian at each iteration to those elements that correspond to grid blocks where a certain minimum saturation
  change has occurred. Other numerical 'tuning' parameters, which may be either
  hard-coded or user-defined are, for example, a maximum allowable number of
  iterations and corresponding shrinkage and growth factors for the time step size
  which are applied depending on whether or not the maximum is reached. Often,
  the values of such parameters are problem-dependent, and some trial and error is
  required to find their optimal values.

3.3.7 Control Aspects*

As mentioned before, the porous-media equations are usually control-affine, i.e.
the controls $\mathbf{u}_k$ enter the equations linearly. An exception is the case when the
inputs consist of a nonlinear combination of ICV settings and bottom-hole pressures; see Sect. 2.3.5.

However, from a control perspective there is a difference between single-phase
and two-phase flow. In the single-phase case, the dynamic response of the
autonomous equations (i.e. without inputs $\mathbf{u}_k$) to a small disturbance (in the form of
an initial condition $\mathbf{x}_0$) from an equilibrium situation (constant pressures in two
horizontal dimensions, or hydrostatic pressures in three dimensions) results in a
return to the equilibrium situation. However, in the two-phase case the saturations
are driven by convection (which is governed by the spatial distribution of the
pressures), and a small disturbance (in the pressures and/or the saturations) from an
equilibrium situation will result in a permanent small change in the saturations.
Correspondingly, the eigenvalues of a continuous-time linear (or linearized) single-phase system have real negative values, whereas only half of the eigenvalues
of a linearized two-phase system are real and negative (for the pressure states)
while the other half are equal to zero (for the saturation states). The two-phase
response is therefore still bounded and non-oscillatory. In other words, the single-
phase system equations are asymptotically stable, whereas the two-phase equations
are *Lyapunov stable*.

Another difference between single-phase and two-phase flow is in the steady-
state behavior of the system. In the single-phase case, steady-state flow may occur
after dampening out of the pressure transients. However, the effect of the convective behavior of the saturations is that there does not exist any non-trivial

two-phase steady-state solution for inputs that result in flow (i.e. that produce a non-hydrostatic pressure gradient). This is because flow produces (very slow) saturation changes as long as there are two mobile phases present (The trivial solution occurs when all mobile oil has been flushed out of the reservoir which effectively makes the reservoir single-phase.) The typical time scale for pressure changes (e.g. defined as the half time for dampening out of an impulsive pressure disturbance in a well) is very small (typically in the order of hours to days) compared to the time for saturation changes to propagate through the entire reservoir (typically in the order of years to decades). Therefore it is usually justified to consider the saturation field to be very slowly time-varying. The pressure response is then governed by linear equations with (very slowly) time-varying coefficients, and, after dampening out of pressure transients resulting from initial conditions, may be considered to be in near-steady state.

3.3.8 Stream Line Simulation*

The key element of the IMPES and sequential simulation schemes is the separate solution of two sets of equations, one for pressures and one for saturations, which are only mildly coupled through the coefficients. A further step can be made by redefining the saturation equations such that they can be expressed as a system of decoupled equations that can be solved independently from each other. The basis for this redefinition is the insight that the saturation equations are mainly convective, or, in other words, that the saturation changes in the reservoir are mainly driven by a velocity field.[14] Here the velocity field refers to the total fluid velocity, i.e. the sum of oil and water velocities. As discussed in Sect. 2.3.6 the velocity field can be 'traced' to generate streamlines, i.e. trajectories of 'virtual' particles traveling through the reservoir. Assuming incompressible flow and a situation where the flow is entirely driven by injection and production wells, the streamlines all start at an injector and end at a producer. The time it takes a particle to travel from the injector to a certain point $\hat{s}$ along its streamline is known as the *time-of-flight* τ which can be expressed as

$$\tau = \int_0^{\hat{s}} \frac{\phi}{v_t} ds \, , \tag{3.107}$$

where $v_t = |\mathbf{v}_t|$ is the magnitude of the total Darcy velocity, ϕ is porosity and s is a coordinate along the streamline starting at the injector. Equation (3.107) can be differentiated with respect to s resulting in the relationship

---

[14] In our case, where we neglected the diffusive effect of capillary pressures, the continuous form of the saturation equation is in fact completely convective. The spatial discretization brings back some numerical diffusion again.

$$\frac{d\tau}{ds} = \frac{\phi}{v_t} ,$$

(3.108)

which can be used to convert expressions in terms of streamline coordinate s to equivalent expressions in terms of time-of-flight τ. In particular, consider the one-dimensional Buckley-Leverett equation (1.73) expressed in streamline coordinate s:

$$v_t \frac{\partial f_w}{\partial S_w} \frac{\partial S_w}{\partial s} + \phi \frac{\partial S_w}{\partial t} = 0 .$$

(3.109)

With the aid of Eq. (3.108) we can express the Buckley-Leverett Eq. (3.109) in terms of time-of-flight coordinate τ as

$$\frac{\partial f_w}{\partial S_w} \frac{\partial S_w}{\partial \tau} + \frac{\partial S_w}{\partial t} = 0 .$$

(3.110)

In analogy to the analytical solution for the Buckley-Leverett equation derived in Sect. 1.4.5, see Eq. (1.93), we can express the solution of Eq. (3.110) as

$$\tau(S_w, t) = \begin{cases} \frac{df_w}{dS_w} t, & S_w^* \leq S_w \leq 1 - S_{or} \\ v_D^* t, & S_{wc} \leq S_w \leq S_w^* \end{cases} .$$

(3.111)

where the dimensionless shock velocity v_D^* is given by Eq. (1.91). Note that the time-of-flight, although expressed in units of time, has taken the role of the spatial coordinate. Expression (3.111) allows us to determine the saturation at a point along a streamline if the corresponding time-of-flight is known. This is a powerful relationship because it is easy to compute the time-of-flight along a streamline once the velocity field has been computed. This, in turn, is a simple step once the pressure field has been computed, as was discussed in Sect. 1.4.12; see Eq. (1.143). The streamlines can then be traced using the expressions given in Sect. 2.3.6, which include the computation of the grid-block travel times $\Delta\tau$, see Eq. (2.180), which can be summed to obtain the time-of-flight. In practice, Eq. (3.110) is usually solved with the aid of finite differences, especially when complicating effects such as compressibility and gravity have to be accounted for. A major advantage of computing the saturations using finite differences along streamlines instead of using finite differences on a conventional spatial grid is an improved stability criterion for explicit time stepping. As discussed in Sect. 3.3.4, the stability condition for explicit Euler integration of the saturation equations is governed by grid blocks with the highest throughput, i.e. those with the smallest spatial dimensions and the highest total velocities. In the numerical simulation of saturations along streamlines it is possible to select the 'grid' size, in terms of time-of-flight increments, independently from the underlying spatial grid, such that much larger time steps can be accommodated. In addition, the stability condition can be determined independently for each streamline, such that for streamlines with low velocities a much larger time step can be chosen than for those with high velocities. Because of this reason, streamline simulation is popular for the fast computation of saturation fields. Streamline simulation becomes particularly attractive in situations where only infrequent updating of the pressure field is required, and therefore also only infrequent repetition of the stream line tracing procedure is needed. The most popular applications involve water flooding with small (or no) compressibility, no or modestly nonlinear relative permeabilities, and fixed wells settings. However the application area is becoming much wider and we refer to Bratvedt, Gimse and Tegnander (1996), Batycky, Blunt and Thiele (1997), King and Datta Gupta (1998) and to the text book of Datta-Gupta and King (2007) for further reading.

3.4 Examples

3.4.1 Example 1 Continued: Stability

3.4.1.1 Stability Limit

We perform a numerical integration of the state equations for Example 1 as defined in Sect. 2.2.2. We choose the input vector as $\mathbf{u}^T = [0.01 \ -0.01]$, which implies that the wells in grid blocks 1 and 6 inject and produce at a rate of 0.01 m^3/s (864 m^3/d, or 5434 bpd). Note that a negative flow rate implies production. If we integrate with an explicit Euler scheme from $t = 0$ until $t = 365 \times 24 \times 3600$ s (i.e. for one year) with a time step of 1 day we obtain the output depicted in Fig. 3.4. The pressure increase at the injector is smaller than the pressure decrease at the producer because of the different permeabilities in the corresponding grid

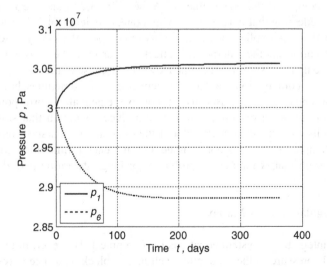

Fig. 3.4 Numerical integration of Example 1 using explicit Euler integration with a time step of 1 day

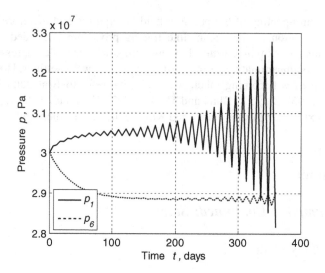

Fig. 3.5 Numerical integration of Example 1 using explicit Euler integration with a time step of 5.8 days, displaying numerically unstable behavior

blocks. If we use a time step of 5.8 days we obtain the spurious result as depicted in Fig. 3.5, because we exceeded the stability limit (the CFL condition) for explicit Euler integration.

3.4.1.2 Singular System Matrix

Earlier we computed the eigenvalues of **A**, see Eq. (3.22), and we found a zero eigenvalue reflecting that **A** is singular with a rank deficiency of 1. In Sect. 3.2.3 it was shown that this implies the impossibility to compute the steady state solution. Here we mention another, numerical, effect. In our case the zero eigenvalue is equal to zero up to 14 significant digits. However, because of the finite precision of the numerical computations, a 'zero' eigenvalue may sometimes have a small positive or negative value. Because a positive eigenvalue corresponds to an exponentially growing response, this may introduce a solution that slowly drifts away from its correct steady-state value if the integration is pursued long enough. For our example the effect is not an issue, but, in general, time integration with a singular system matrix may cause problems for long integration periods.

3.4.1.3 Regular System Matrix

We again integrate the system equations for Example 1. However, in this case the bottom-hole pressure of the production well in grid block 6 has been prescribed as $p_{well,6} = 26.00 \times 10^6$ Pa (3771 psi), while the injection rate in block 1 remains

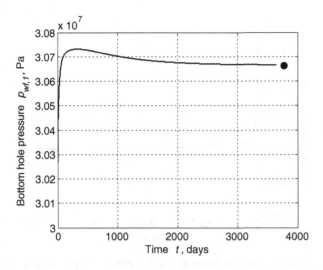

Fig. 3.6 Numerical integration of Example 1 with a prescribed pressure in grid block 6. The figure shows the bottom-hole pressure in the injection well in grid block 1. The *dot* represents the steady-state result

fixed at $q_1 = 0.01$ m^3/s (864 m^3/d, 5434 bpd). Figures 3.6 and 3.7 give plots of the output variables versus time. The solid dots represent the steady-state results computed with the aid of Eqs. (3.40) and (3.41). Note that we needed to integrate for a period of around 3000 days before the pressure in the injection well approached the steady state result closely (3.06668 MPa after 10 years versus 3.06655 MPa fully steady-state).

3.4.2 Example 2 Continued: Mobility Effects

For two-phase flow we present examples obtained with a simple in-house simulator. We start with the forward simulation of Example 2, i.e. the same six-block model that was used to illustrate single-phase flow behavior, but with additional reservoir and fluid properties as given earlier in Sect. 1.4.4. We choose the operating conditions such that water is injected at a constant rate of 0.01 m^3/s in grid block 1, while liquid is produced at a constant well-bore pressure of 20 MPa in grid block 6. Because of the very small size of the model we can use explicit Euler integration, and Fig. 3.8 depicts the output for a simulation time of 10000 days (approximately 27 years). The total injected water volume is 8.64×10^6 m^3 which amounts to 2.4 times the total volume of moveable oil.[15]

---

[15] The moveable oil volume is equal to the pore volume times $(1 - S_{wc} - S_{or})$.

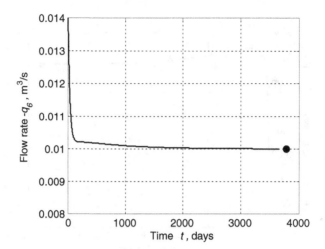

Fig. 3.7 Numerical integration of Example 1 with a prescribed pressure in grid block 6. The figure shows the absolute value of the flow rate in the production well in grid block 6. The *dot* represents the steady-state result

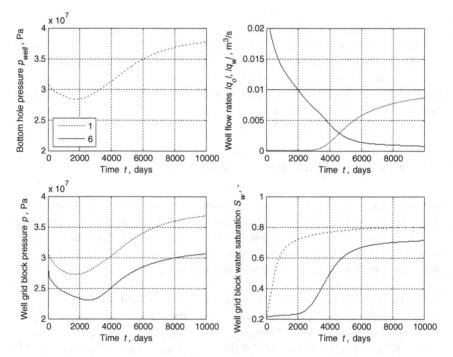

Fig. 3.8 Results for numerical integration of Example 2 with a prescribed water injection flow rate in grid block 1 and a prescribed pressure in grid block 6. The *solid* and *dotted lines* in the *top right* figure represent the oil and water production rates in grid block 6; the *solid horizontal line* represents the injection water rate in grid block 1. In the other three figures the *dashed* and *solid lines* refer to results for the injection and production wells in grid blocks 1 and 6 respectively

In the top-right figure it can be seen that water breakthrough in the producer occurs after about 3000 days.

The bottom right figure displays the water saturations in the well grid blocks. Starting from connate-water saturation (0.2) they approach a value of one minus residual-oil saturation (0.8) more or less gradually. The relatively large grid blocks, compared to the total domain, cause a large amount of numerical diffusion. The two figures at the left illustrate that the well-bore pressure in the injector, and the grid-block pressure in both well grid blocks behave non-monotonously. Initially they drop because the prescribed pressure in the producer (20 MPa) is considerably below the initial reservoir pressure (30 MPa). However, when the oil-water front approaches the producer the total relative mobility in the producer grid block decreases because of the nonlinear saturation dependency of the relative permeabilities (see the dotted line in Fig. 1.4) and because of the viscosity difference between oil and water (0.5×10^{-3} versus 1.0×10^{-3} Pa s respectively). The resulting increased pressure drop over the near well-bore area of the producer, as represented in the well model, results in an increase in pressure in the entire reservoir. Note that if the injector had been operated at a prescribed pressure instead of at a prescribed rate, this mobility effect would have resulted in a drop in total production rate instead of an increase in reservoir pressure.

3.4.3 Example 3 Continued: Well Constraints

Next we consider the forward simulation of Example 3 which was described in Sect. 1.4.9. The initial operating constraint for the injector is specified as a prescribed rate of 0.002 m^3/s (1087 bbl/d) with a maximum bottom-hole pressure constraint equal to 35 MPa (5076 psi) which is 5 MPa (725 psi) above the initial reservoir pressure. The initial operating constraint for the producers is a prescribed pressure of 25 MPa (3626 psi), i.e. 5 MPa (725 psi) below the initial reservoir pressure, with a maximum flow rate per well equal to -0.001 m^3/s (-543 bbl/d). The initial water saturation is equal to the connate-water saturation (0.2), and the total moveable oil volume is 16700 m^3 (1.05×10^6 bbl). We use implicit Euler integration with Newton-Raphson iteration, and a variable time step with maximum allowed pressure and saturation changes of 1×10^6 Pa and 0.2 respectively, target changes equal to 90 % of these values, and a maximum time step of 30 days. Because of the relatively small problem size (882 states) the linear system of equations within each Newton-Raphson iteration is solved with the aid of a direct solver. Figure 3.9 displays the output for a simulation time of 1500 days (approximately 4.1 years). In the top left figure it can be seen that the bottom-hole pressure in the producers stays at its prescribed pressure of 25 MPa (3626 psi) during the entire period and in the top right figure it can be verified that the production rates in the producers never exceed the maximum allowed flow rate of

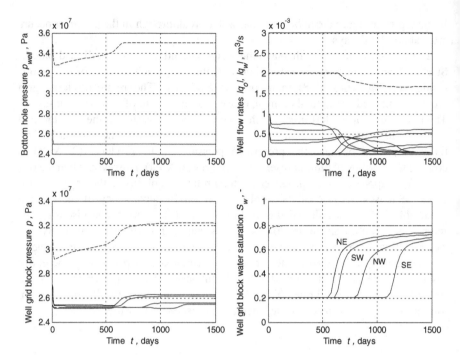

Fig. 3.9 Well production data for Example 3. The *solid* and *dotted lines* in the *top right* figure represent the oil and water production rates in the producers; the *dashed line* represents the injection water rate. In the other three figures the four *solid lines* refer to the four producers, and the *dashed line* to the injector. The letters NE refer to the North-East (*top right*) producer, the letters SW to the South-West (*bottom left* producer), and so on

−0.001 m³/s (−543 bbl/d). The injector, however runs against its pressure constraint after approximately 700 days: the top left figure shows that until that time the pressure stays below the constraint of 35 MPa (5076 psi) and, correspondingly, the top right figure shows a steady injection rate of 0.002 m³/s (1087 bbl/d). After reaching the constraint the injector is effectively operating at a prescribed bottom-hole pressure of 35 MPa (5076 psi) with a maximum rate constraint equal to 0.002 m³/s (1087 bbl/d). This new constraint is not reached anymore in the remaining time, so no further constraint switches occur. The bottom right figure displays the water saturations in the well grid blocks and it can clearly be seen that there is a considerable difference in arrival time of the water front in the four producers. The same effect can be observed in the oil and water well flow rates depicted in the top right figure.

The bottom left figure displays the grid-block pressures, and, just as in Example 2, displays a clear mobility effect when the water front reaches the producers. The effect is an increase in pressure which rapidly spreads through the entire reservoir and which therefore results in the injector reaching its maximum bottom-hole

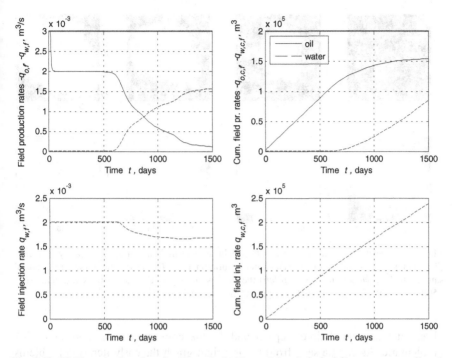

Fig. 3.10 Field-production data for Example 3. The *solid* and *dashed lines* represent oil and water rates respectively

constraint as was described above. The corresponding injection and production field rates, i.e. the sums of the well rates, have been depicted in Fig. 3.10. They show a typical oil production plateau followed by a rapid decline and a simultaneous increase in water production. The decline in oil production is not only caused by the increase in water cut in the producers but also by the inability of the injector to maintain its maximum rate because it has run into its pressure constraint. Figure 3.11 depicts 8 snapshots of the water saturation at different moments in time. The effect of the high permeable streak can clearly be seen: water breakthrough occurs in the North-East corner first, followed by the South-West, North-West and South-East corners, a sequence that is in line with the saturation curves in Fig. 3.9 (bottom right).

3.4.4 Example 3 Continued: Time-Stepping Statistics

Figures 3.12 displays several numerical parameters that give an indication of the functioning of the implicit variable-time step integration process. The top left graph displays the number of Newton iterations per time step, and it can be

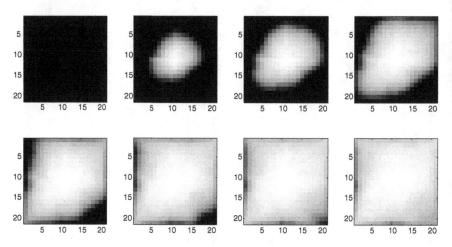

Fig. 3.11 Snapshots of the water saturation field at time intervals of approximately 214 days (7 months) on a scale from connate-water saturation (0.2, *black*) to one minus residual-oil saturation (0.8, *white*). Initially (*top left*) the field is entirely at connate-water saturation. After 1500 days (*bottom right*) the field is approaching residual-oil saturation

observed that around time steps 3 and 17 the convergence became somewhat problematic. As can be seen from the top right graph the early iteration problems are related to a high number of constraint violations per time step. In response, the time-stepping algorithm repeats the Newton-Raphson iterations with a progressively reduced time step size until convergence without constraint violation is reached. The targets for grid-block pressure and saturation changes were specified as 90 % of 1×10^6 Pa and 0.2 respectively, with a maximum growth factor of 0.7 per time step, and a maximum time step size of 30 days. It can be seen from the two graphs in the middle row of Fig. 3.12 that the target values were never met. This is because initially the time step size was occasionally reduced to obtain convergence, and as of time step 48 because the maximum allowed time step size was reached; see also the bottom left graph of Fig. 3.12 which displays the size of each time step. The bottom-right graph of Fig. 3.12 displays the mass-balance error ε_m during each time step k defined as

$$\varepsilon_{m,k} \triangleq \frac{m_k - m_0}{m_0} = \frac{m_{1,k} + m_{2,k} + m_{3,k} + m_{4,k} - m_{5,k}}{m_{1,0} + m_{3,0}} - 1 \, , \qquad (3.112)$$

where m_1 is the mass of oil in all grid blocks, m_2 the cumulative mass of produced oil, m_3 the mass of water in all grid blocks, m_4 the cumulative mass of produced water, and m_5 the cumulative mass of injected water. A small mass-balance error develops during the simulation because we do not use a fully mass-conservative formulation, but the maximum error never exceeds 1.5 % in this example, while at the end of the simulation it is less than 0.4 %.

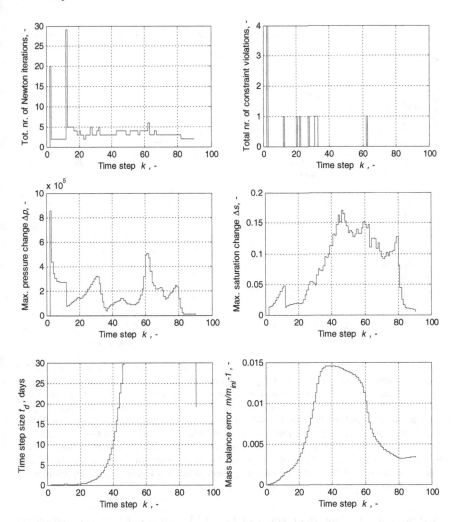

Fig. 3.12 Numerical parameters. Total number of Newton-Raphson iterations per time step (*top left*), total number of constraint violations of bottom-hole pressure or total well rate per time step (*top right*), maximum grid-block pressure change per time step (*middle left*), maximum grid-block saturation change per time step (*middle right*), time step size (*bottom left*), and mass-balance error (*bottom right*)

3.4.5 Example 3 Continued: System Energy*

Figure 3.13 illustrates the power balance for Example 3, where the various contributions have been computed with the aid of Eq. (2.183) of Sect. 2.3.7. The top left figure illustrates that during a brief initial period potential energy is released from the reservoir through oil flow, but that rapidly an equilibrium is established during which the total amount of potential energy stored stays nearly constant and

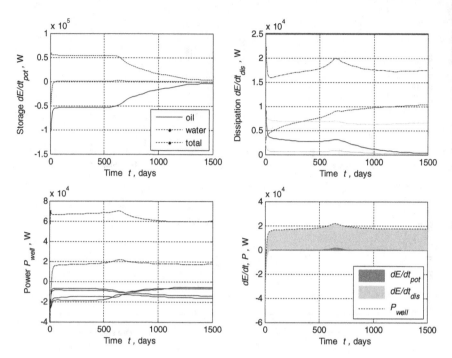

Fig. 3.13   Power balance for Example 3. The *solid, dotted* and *dash-dotted lines* in the *top left* figure represent the oil, water and total potential energy storage rates respectively. The *solid, dotted* and *dash-dotted lines* in the *top right* figure represent the oil, water and total dissipation rates respectively, where the *black solid* and *dotted lines* refer to dissipation through the grid-block boundaries, and the light *gray solid* and *dotted lines* to dissipation in the near-well-bore area. In the *bottom-left* figure the four *solid lines* refer to the power flow in four producers, the *dashed line* to the power flow in the injector, and the *dash-dotted line* to the total power flow in the wells. The *bottom-right* figure illustrates that the total power flow in the wells (*dash-dotted line*) equals the sum of the total potential energy storage rate (dark gray area; hardly visible) and the total dissipation rate (*light gray area*)

close to zero. The top-right graph in Fig. 3.13 displays the energy dissipation caused by oil and water flow through the grid-block boundaries and in the near-well-bore area. Note that the ratio between near-well-bore and grid-block losses would become progressively smaller with decreasing grid size. Not surprisingly, the dissipation caused by oil flow reduces with time while the dissipation caused by water flow increases, corresponding to the increasing water-oil ratio of the produced reservoir fluids. The effects of relative permeabilities are visible in the small increase halfway the producing reservoir life when the water front reaches the producers. The bottom-left graph displays the power flow in the wells. It can be seen that there is an influx of energy through the injector and an outflow through the producers, resulting in a near-constant net influx of approximately 17 kW with a small peak to just above 20 kW.

Note that these values do not take into account the effect of elevation-related potential energy, which would change the situation. E.g. if we would assume that the reservoir were located at a depth of $d = 3000$ meter, and were initially hydrostatically pressured with oil and water densities $\rho_w = 850$ kg/m^3 and $\rho_w = 1000$ kg/m^3, and an acceleration of gravity $g = 9.81$ m/s^2, then the elevation-related energy in a totally oil-filled well would be equal to

$$E_{lift} = (\rho_w - \rho_o)gd = (1000 - 850) \times 9.81 \times 3000 = 4,414,500 \text{ J} . \quad (3.113)$$

Assuming a well-bore radius $r_{well} = 0.114$ m, the total well volume would be

$$V_{well} = \pi r_{well}^2 d = 3.14 \times 0.114^2 \times 3000 = 122.4 \text{ m}^3 , \quad (3.114)$$

such that with an average production rate of $q_o = 0.5$ m^3/s the well-bore contents would be emptied in

$$t_{lift} = \frac{V_{well}}{q_o} = \frac{122.4}{0.5} = 245 \text{ s} . \quad (3.115)$$

The elevation-related lifting power of a completely oil-filled well would then be

$$P_{lift} = \frac{E_{lift}}{t_{lift}} = \frac{4,414,500}{245} = 18018 \text{ W} . \quad (3.116)$$

This simple analysis does not even take into account the additional lift effect of gas escaping from oil in the well bore. However, it should be noted that if we would not inject water, a very rapid reduction in reservoir pressure would occur and soon after start of production the wells would stop flowing.

References

Aziz K, Settari A (1979) Petroleum reservoir simulation. Applied Science Publishers, London

Batycky RP, Blunt MJ, Thiele MR (1997) A 3D field-scale streamline-based reservoir simulator. SPE Reserv Eng 12(4):246–254. doi:10.2118/36726-PA

Boyce W, Di Prima RC (2005) Elementary differential equations and boundary value problems, 8th edn. Wiley, New York

Bratvedt F, Gimse T, Tegnander C (1996) Streamline computations for porous media flow including gravity. Transp Porous Media 25(1):63–78. doi:10.1007/BF00141262

Chen Z, Huan G, Ma Y (2006) Computational methods for multiphase flows in porous media. SIAM, Philadelphia

Datta-Gupta A, King MJ (2007) Streamline simulation: theory and practice, SPE Textbook Series, 11. SPE, Richardson

King MJ, Datta-Gupta A (1998) Streamline simulation: a current perspective. In Situ 22(1):91–140

Luenberger DG (1979) Introduction to dynamic systems. Wiley, New York

Moler C, Van Loan C (1978) Nineteen dubious ways to compute the exponential of a matrix. SIAM Rev 20(4):801–836

Index

J. D. Jansen, *A Systems Description of Flow Through Porous Media*,
SpringerBriefs in Earth Sciences, DOI: 10.1007/978-3-319-00260-6,
© The Author(s) 2013